GW01458507

THE PHYSIOLOGY OF TROPICAL
CROP PRODUCTION

To
Evaporation and Environment
and
the field worker

The Physiology of Tropical Crop Production

by

G. R. Squire

C·A·B International
for
The Overseas Development Administration

C·A·B International
Wallingford
Oxon OX10 8DE
UK

Tel: Wallingford (0491) 832111
Telex: 847964 (COMAGG G)
Telecom Gold/Dialcom: 84: CAU001
Fax: (0491) 833508

First printed 1990
Reprinted 1993

British Library Cataloguing in Publication Data

Squire, G. R. (Geoffrey R.)
    The physiology of tropical crop production.
    1. Tropical regions. Crops. Physiology
    I. Title II. CAB International
    631.50913

ISBN 0-85198-677-3

Typeset by Enset Photosetting, Midsomer Norton, Bath
Printed and bound in the UK

Contents

Preface

The aim of this book is to provide an introduction to the principles of crop physiology in the tropics. A research grant from the Overseas Development Administration (Natural Resources and Environment Department) of the UK, enabled myself and an assistant, Mr David Amos, to work part-time on analysis and writing for 12 months in 1988 and 1989. The grant also paid for 500 copies to be printed and distributed to research stations and universities in the developing countries of the tropics.

The idea of the volume, and its structure and content, owe much to an earlier project based at the University of Nottingham, also funded by the ODA (see Introductory Remarks), and I am pleased to acknowledge the many direct and indirect contributions by my colleagues in that project. Access to journals and books was generously provided by several libraries around Aberdeen and Edinburgh.

G. R. S.

Introductory Remarks

This book examines the way the physiological processes of tropical crops are governed by the elements of the physical environment, chiefly solar radiation, temperature, humidity, soil water and nutrients. The treatment of the subject developed from work funded by the UK Overseas Development Administration at the University of Nottingham, Sutton Bonington, in which the author participated. This work examined the physiological control of yield by environmental factors, and concentrated on the effects of temperature and drought, omitting any consideration of nutrients and other soil factors (since these would have required a major project themselves). The work was confined to the responses of representative species – two cereals, pearl millet and sorghum, and one legume, groundnut. Moreover, the emphasis was on those physiological processes that were among the least understood, and were receiving scant attention elsewhere.

The report of the work* included an account of the principles governing the responses of those crops to temperature and drought. The present volume is based on that account, but attempts to achieve a more rounded synthesis. A few paragraphs of the report, and a few tables and illustrations are included here with little change, but the rest of the text and most illustrations are new. More physiological processes and environmental factors are examined, with examples from the main groups of the tropical crops.

The scope of the material for inclusion was extended to a much wider literature within the subject of crop physiology. Much of this new material however, was obtained from papers and reports of other projects funded by the Overseas Development Administration. Some of these were intended to provide fundamental knowledge of plant/weather relations, while others, though with practical aims, revealed much about the control of physiological processes. Among the species examined in this work were sorghum in Botswana, pigeon pea in Trinidad, sugar cane in Mauritius, maize in western

---

Microclimatology in Tropical Agriculture (2 vols) (1987) Overseas Development Administration (Natural Resources and Environment Department), London.

Kenya, tea in several parts of Africa, barley in mediterranean-type climates, rice in Sri Lanka and several subsistence crops in the Yemen. There was also work in the laboratory, for example on the control of development by temperature and photoperiod in legumes, and on the biochemical basis of high-temperature tolerance in dryland cereals.

The amount of relevant information is immense, and to keep the volume to a workable size, the inclusion of material has been selective, on four counts. First, the examples are taken largely from field work. Research in controlled environments is included to illustrate principles, or to distinguish the effects of two or more environmental variables – for example, temperature and photoperiod on flowering, or high temperature and dry air on leaf expansion. But responses in controlled environments, especially to drought and nutrients, can be misleading, and may be irrelevant when root systems are severely confined.

Secondly, levels of organization are considered from the organ or structure, such as the leaf lamina, root branch, grain, pod and fruit, to the crop stand consisting of many individuals of one or more species. There is little description of cellular physiology or gas exchange at one end, or farming systems at the other, though both are linked by the processes examined here.

Thirdly, the account concentrates on a few species representing the main groups of tropical crops. This is to achieve continuity between the chapters. The cultivars of pearl millet and groundnut, described in *Microclimatology in Tropical Agriculture*, are used as the main thread. Other cereals frequently referred to include maize, sorghum and rice, and other legumes, pigeon pea and cowpea. Cassava represents the root and tuber crops. Tree crops are included since their physiological traits contrast with those of the herbs: oil palm and tea are the main representatives, since more is known of their environmental physiology than of most other fruit and beverage crops.

Finally, the field research is selected largely from the developing countries of the tropics. This excludes much fine work in the warmer regions of North America and Australia, a necessity the author regrets.

Information in tables and illustrations

The material in tables and graphs used to complement the text is taken from the published literature, though with few exceptions the information is not presented in its original form. Mostly, the data have been re-worked, and many graphs consist of information from more than one source. This approach sometimes has satisfying results, as when data collected by different scientists, working many years and miles apart, are combined to illustrate a principle. Where the original data allow, the sampling variability of physiological attributes is indicated by standard errors.

References

Numbers in square brackets in the text, e.g. [9] , indicate that information relevant to the point can be found in the lists at the end of the Chapter. Many of these 'chapter references' show that the source of the information is in one of the papers or reports of original research listed in a separate section after the final chapter, conventionally, in alphabetical order of the authors' names. Suggestions of relevant books and reviews for further reading are also given at the end of each chapter.

General reading

This treatment of crop physiology touches on plant physiology, micro-climatology and soil science, and assumes some familiarity with these subjects. Milthorpe and Moorby's book provides a sound basis, and surveys the field up to the late 1970s, with examples mainly of temperate species; reference to many of the standard works in crop physiology can be found in the lists of further reading at the end of its chapters. The new, much revised *Russell* is indispensable. G. C. Evans' treatise gives a detailed account of growth analysis and points to some of the classical works in the subject. J. L. Monteith's essay is a starting point for some recent analyses.

There are several many-authored volumes dealing either exclusively or partly with tropical species. The third, fourth and seventh references all start with a general account of crop physiology, and consist mostly of chapters on individual species. Several chapters of Eastin *et al.* have become standard works that have hardly dated.

Eastin, J. D., Haskins, F. A., Sullivan, C. Y. and Van Bavel, C. H. M. (eds) (1969) *Physiological Aspects of Crop Yield*. American Society of Agronomy, Wisconsin, USA.

Evans, G. C. (1972) *Quantitative Analysis of Plant Growth*. Blackwell Scientific Publications, Oxford.

Evans, L. T. (ed.) (1975) *Crop Physiology – Some Case Histories*. Cambridge University Press.

Goldsworthy, P. R. and Fisher, N. M. (eds) (1984) *The Physiology of Tropical Field Crops*. Wiley-Interscience, London. The chapter by M. D. Dennett (pp. 1–38) gives a thorough introduction to the tropical climate.

Milthorpe, F. L. and Moorby, J. (1979) *An Introduction to Crop Physiology* (2nd edition). Cambridge University Press.

Monteith, J. L. (1981) Climatic variation and the growth of crops. *Quarterly Journal of the Royal Meteorological Society*, **107**, 749–74.

Norman, M. J. T., Pearson, C. J. and Searle, P. G. E. (eds) (1984) *The Ecology of Tropical Food Crops*. Cambridge University Press.

Wild, A. (ed.) (1988) *Russell's Soil Conditions and Plant Growth* (11th edition), Longman, London.

List of Symbols

Commonly used symbols with typical units where appropriate.

D   atmospheric saturation water vapour pressure deficit (kPa), shortened to saturation deficit.

D′ the leaf-to-air saturation deficit (kPa), which is the same as D if leaf and air temperatures are the same.

E_o potential evaporation rate (mm d^{-1}), estimated from a formula.

E_a actual evaporation rate from all sources in a stand (mm d^{-1}); components –
     E_i water intercepted by foliage,
     E_s water from the soil surface,
     E_t transpiration through plants.

E_l evaporation rate per unit leaf area (g m^{-2} h^{-1}).

f   the fraction of the incoming solar radiation intercepted by a plant canopy (fractional interception); $\bar{f}$ is a mean fractional interception over a specified period of time.

g_c canopy conductance, determined as the sum of the conductances (to water vapour) of different types or layers of foliage weighted by the corresponding leaf area indices (cm s^{-1}).

I   inflow of water per unit length of root per unit time (g m^{-1} d^{-1}).

k   the extinction coefficient of a canopy (relating L to f).

l_v root length density, the length of root per unit volume of soil (cm cm^{-3}).

L   leaf area index (leaf area per unit field area).

N_p plant population density (m^{-2} or ha^{-1}).

p   partition factor of a structure – usually determined as the slope of the linear regression of the dry mass of the structure on the dry mass of the whole plant, from a series of samples taken during the growth of the structure (dry mass expressed per plant or per unit field area).

p′ the slope of the linear regression of the dry mass of a structure (usually the reproductive) on the dry mass of the whole plant *at final harvest*, over a range of population density (dry mass expressed per plant).

P   photoperiod (h).

R   length of root per unit field area (km m^{-2}).

S the daily incoming solar radiation (total, as distinct from photosynthetically active) ($mJ \, m^{-2} \, d^{-1}$).

S$_i$ the amount of S intercepted by a plant canopy, i.e. reflected and absorbed ($mJ \, m^{-2} \, d^{-1}$).

t the time for a development process to be completed, usually analysed as the reciprocal 1/t (the developmental rate); sometimes identified by a subscript to denote a specific process, e.g. **t$_s$** for initiation of a series of leaf primordia, **t$_w$** for growth in weight of a structure (d).

T temperature ($°C$).

T$_b$ the base temperature for a process of development or expansion (at and below which the process does not proceed); sometimes further identified as that controlling a rate of expansion, **T$_{br}$**, or a developmental duration, **T$_{bd}$** ($°C$).

T$_o$ the optimum temperature for rate of development or expansion ($°C$).

T$_c$ the ceiling temperature for development or expansion (at and above which the process does not proceed); synonymous with **T$_m$** the maximum temperature, used by some authors ($°C$).

W the dry weight or mass of a plant or stand (g or $g \, m^{-2}$), identified by **W$_t$** in Chapter 5 only, to distinguish it from:
  W$_s$ the dry mass of a specified structure,
  W$_o$ the dry mass of the plant or stand at the start of growth of a specified structure.

W$_v$ the apparent minimum vegetative dry mass of a plant, determined as the intercept on the horizontal axis of a linear regression of reproductive on total dry mass per plant obtained over a range of plant population density.

ε_s the dry-matter/intercepted-radiation ratio – the dry mass of plant tissue (usually above-ground only) accumulated over a period of time, divided by the amount of *solar* radiation intercepted by the foliage over the same period; values usually quoted are for total solar radiation, unless identified as for photosynthetically active radiation ($g \, MJ^{-1}$).

ε_n the dry-matter/nutrient ratio – the dry mass of plant tissue (usually above-ground only) divided by the corresponding mass of a specified *nutrient* in the tissue ($g \, g^{-1}$).

ε_w the dry-matter/transpired-water ratio – the dry mass of plant tissue (usually above-ground only) accumulated over a period of time, divided by the corresponding mass of transpired *water* ($g \, kg^{-1}$).

Γ rate of growth of a whole plant ($g \, d^{-1}$), or stand per unit field area ($g \, m^{-2} \, d^{-1}$).

Θ accumulated thermal time (time-temperature integral above a specified base temperature), not necessarily for a specified process, but usually over a series of processes.

Θ_1 the thermal duration (time-temperature integral above a specified base temperature) for a specified developmental process, when temperature is between T_b and T_o; usually with subscript, e.g.
  Θ_{li} for initiation of leaf primordium,
  Θ_{le} for expansion of a leaf lamina (°Cd).

Θ_2 the corresponding thermal duration when temperature is between T_o and T_c; rarely measured except for germination (°Cd).

Ψ_1 leaf water potential, also:
  Ψ_s the corresponding solute potential,
  Ψ_p turgor potential.

ρ thermal rate of linear extension, e.g. of leaf laminae (mm (°C h)$^{-1}$).

Chapter One

Control of Development

When a plant has germinated, it initiates primordia. Vegetative primordia expand to form a canopy and root system, which, by intercepting radiation and extracting water and nutrients from the soil, produce a store of dry matter and nutrients. While the plant is vegetative, this store is used to generate a larger canopy or more extensive root system. If the plant becomes reproductive, some of the store is allocated to expanding flowers and fruits (Fig. 1.1).

The physiological processes controlling the sequences of vegetative and reproductive growth may be measured in terms of morphological attributes such as length, area and dry mass, or of integrals of an environmental factor such as (transpired) water or (intercepted) solar radiation. The change in time of these attributes can be examined in terms of a *duration* and a mean *rate* (Fig. 1.2). The principles governing duration are common to most types of process and are considered in this chapter. Subsequent chapters examine the characteristics of rate and specific aspects of duration for three types of process: expansion of the canopy and the root system; storage of dry matter in terms of the limiting resource – solar radiation, water or nutrients; and partition of the store of dry matter between the various organs, specially those harvested as the economic yield.

Principles of development

The development of a plant from germination to maturity can be considered as a series of discrete periods, each identified by a process of change in the structure, size or mass of specific organs. Such periods are those required for germination of seeds and emergence of seedlings; for initiation of primordia, whether of a leaf, root, tiller or spikelet; for expansion and for growth in mass of the canopy, stem and root system; and for flowering and filling of panicles. These periods are either determinate or indeterminate. The determinate periods begin at a particular time in relation to a specific event (e.g. sowing) and end, with the exception of the final period, before the crop is

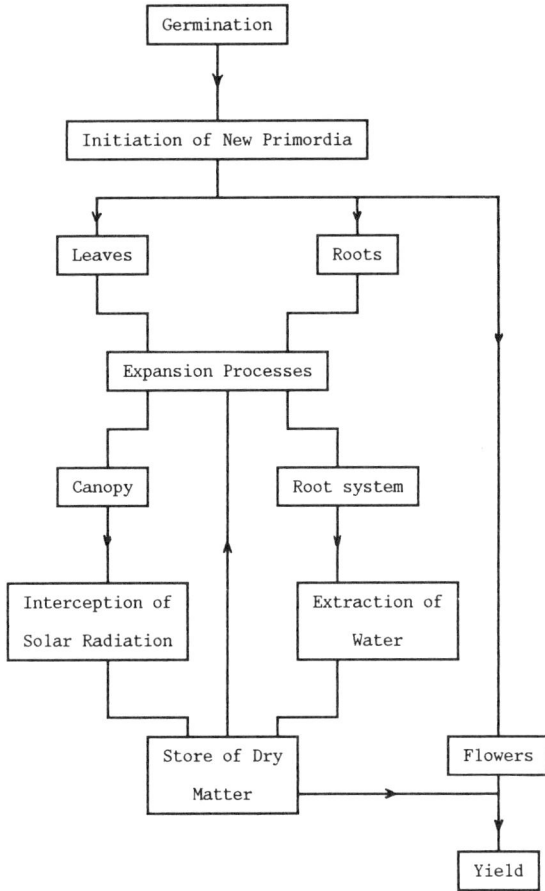

Fig. 1.1. Flow diagram of the main processes in development and growth of a crop.

harvested. The indeterminate periods have a defined beginning, but may continue for many months or years.

Most genotypes can be classified as either determinate or indeterminate, though some have processes which have both determinate and indeterminate properties. In cereals, for example, the developmental period associated with growth in mass of stems is usually similar in duration to that for extension of stems, which is determinate; but when there is a surplus of assimilate, stems continue to accumulate dry matter even when they cease expanding. In many herbaceous legumes, the periods between sowing and the time of first flower or pod, and between successive flowers or pods, and therefore for the production of a given number of flowers or pods, are determinate; but the process of reproduction, once begun, is indeterminate and usually ceases

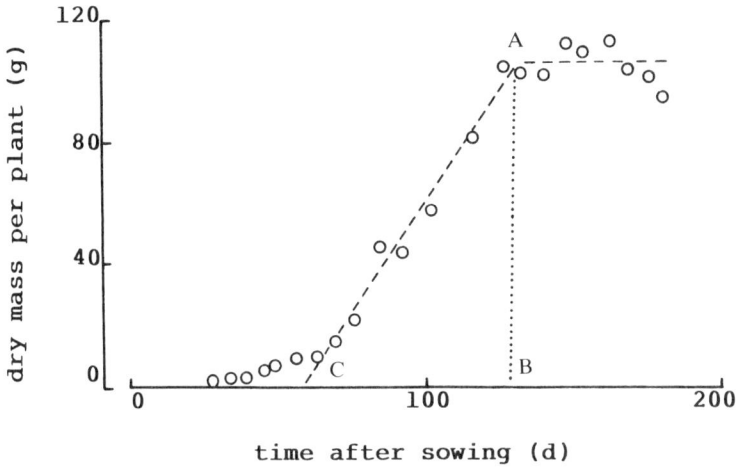

Fig. 1.2. Rate and duration of a process. At the level of the stand, many processes can be examined in terms of a mean rate (AB/BC) and an effective duration (CB). This example is the increase in vegetative dry mass of the pigeon pea cultivar ICP-1, from the data of Sheldrake and Narayanan, 1979.

only when the crop is harvested, or when the canopy is destroyed by disease or drought.

Whether determinate or indeterminate, the development of a structure relies to varying degrees on expansion and growth in mass. Some processes, such as the initiation of successive leaf primordia, are associated with only very small changes in size or mass; others, such as flowering in cereals, which follows expansion of the stem and panicle, can only proceed after much investment in mass. Nevertheless, in many species, the course of development tends to proceed somewhat independently of the rates of expansion and growth.

Rate and duration, population and percentile

Processes of development, as defined here, are discrete. A seed is either not germinated or germinated; a leaf primordium invisible or visible; and a leaf expanding or expanded. Therefore they are defined in terms of time, not length, area, volume or mass. The duration of a developmental process (t) is usually measured between events that are detected by eye. Such durations are those between wetting a seed and the subsequent elongation of the radicle to a defined length, and between fertilization of a flower and the maturity of grain, pod or fruit. The *rate of development* is simply the reciprocal of this duration, and is here identified as 1/t. (So, for example, a developmental rate expressed as 0.1 d^{-1}, indicates that development of the process progresses at a rate of 0.1 or 10% per day, and is complete ten days after it began.)

Most practical estimates of t are made for a sample of individuals taken from a population. As expected, t varies throughout the sample, and a plot of cumulative development (X) on time usually appears as in Figure 1.3. Some members of a population may fail to develop, so the maximum proportion that survives (X_m) is usually less than 1. An important characteristic of such a plot is the *spread* of time between the completion of development for the most and least precocious individuals. (This spread can be approximated by a cumulative normal, or other statistical distribution.) For processes such as germination, the spread is large compared with the mean duration of development, so a given development rate must be identified by a corresponding fraction or percentile of the population[1]. The durations for two such percentiles are shown on Figure 1.3. Others may be chosen for specific purposes. Developmental rates of many processes are measured for X = 0.5.

The following analysis of development concentrates on factors controlling survival and developmental rate. Survival during germination and emergence is particularly sensitive to high temperature and soil dryness and, at other stages, to dryness and plant population density. For surviving percentiles, developmental rate is most generally governed by temperature and photoperiod.

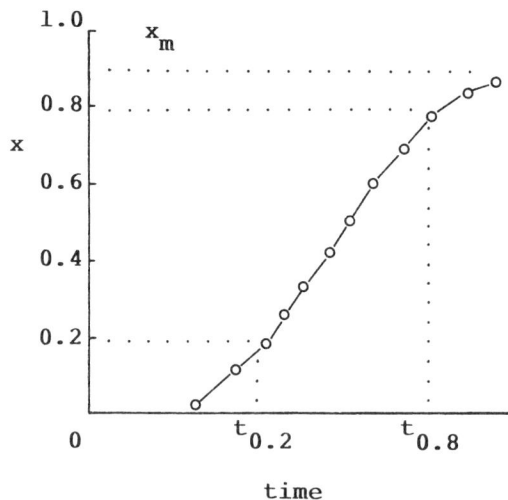

Fig. 1.3. Development and time. Representation of cumulative development (X) of a population in relation to time (t) from the beginning of a relevant event. The duration of development is shown for two subsets of the population, X = 0.2 and X = 0.8. X_m represents the fraction of the population that completes development.

Temperature and development

Survival

The chances that an individual will not survive to complete a developmental phase are greatest during germination, emergence and the early phases of leaf and root initiation. Generally, for healthy seeds that have not been stored for a long time, X_m during these stages is at least 0.9 over a wide range of temperature, below and above which X_m decreases sharply to zero (e.g. Fig. 1.4a). Most seeds of tropical species survive between 15–40°C, though few survive below 10°C or above 50°C. These ranges are much wider for certain sub-tropical species such as chickpea, that are nevertheless widely grown in the tropics[1].

Tolerance of extreme temperature

Tropical species are more likely to experience temperatures near the upper limit of their range than the lower, particularly during seedling emergence in drier regions. There is evidence that the degree of tolerance to such temperatures is influenced by the temperature previously experienced by the germinating seed or the seedling. For example, if the germinating seeds of many tropical species are suddenly exposed to a temperature of 45–50°C, further development is delayed or prevented. The high temperature appears to restrict or inhibit protein synthesis, but probably also affects other cellular processes. The restriction is greater if the high temperature is experienced early, during imbibition, than when germination has already started. If, however, the seeds are first given a very brief exposure to a high but not lethal temperature, or if temperature is raised slowly to the high value, the seed is better able to tolerate the high temperature. This acquired 'thermal tolerance' is attributed to the formation of a specific type of protein (sometimes termed heat-shock protein) that makes the tissue less sensitive to the high temperature. These proteins are either not produced, or at least produced slowly, during the early stages of imbibition, which might explain the greater sensitivity of germination at that stage[2].

Much of this work on high-temperature responses has been with germinating seeds in the laboratory. The relevance of the responses in the field has still to be determined. However, the linkages between these biochemical attributes and the ability of a genotype to survive offer a possible means of screening genotypes for high-temperature tolerance. For example, in one experiment, protein synthesis in the embryo during germination was impaired less in tolerant than in susceptible genotypes. In another, tolerants produced heat-shock protein before susceptibles[2].

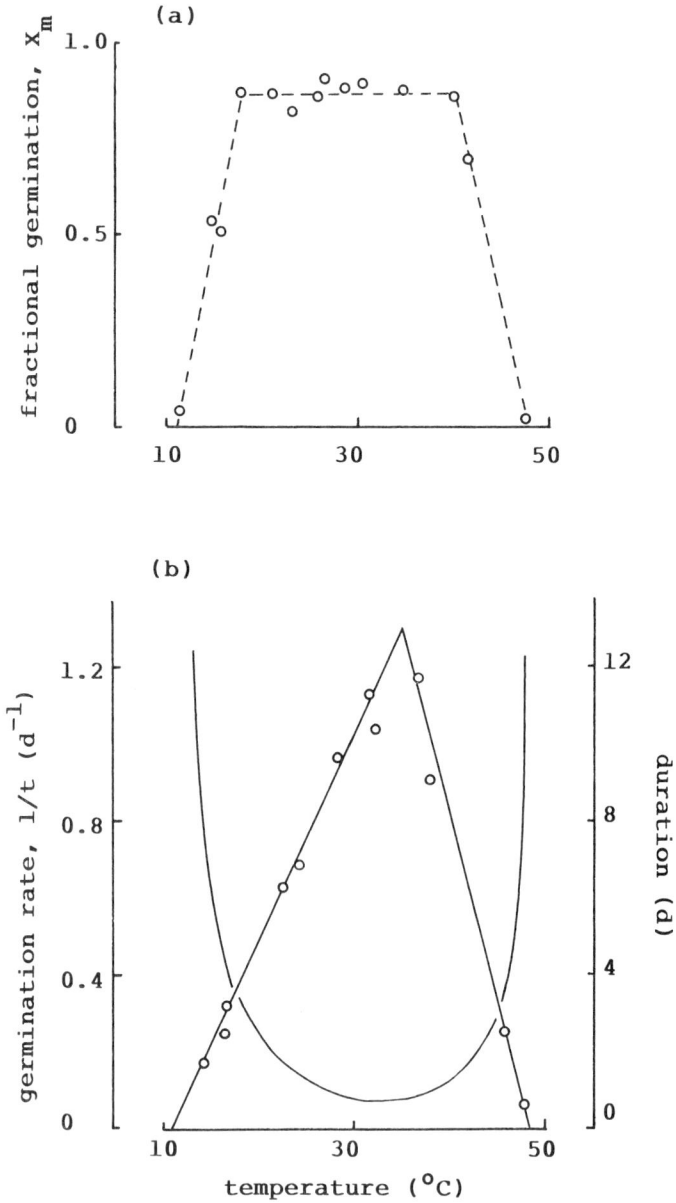

Fig. 1.4. Germination and temperature. For seeds of pearl millet obtained at a food market in the Yemen: (a) the fraction of the sample that germinated; (b) the rate of germination (○), defined as $1/t$ to $X = 0.5$ (see Figure 1.3), and the corresponding duration, $= 1/rate$ (————). Source: Garcia-Huidobro *et al.*, 1982a.

Rate of development

The duration (t) of a process usually decreases with rise in temperature to a minimum at an optimum temperature, and increases with further rise in temperature above the optimum. The responses on either side of the optimum are not linear: a given change of temperature has a greater effect further from the optimum. However, the rate of development (1/t) increases more or less linearly with rise in temperature (T) above a base or threshold temperature, T_b, at which duration is infinite, to an optimum temperature (T_o), at which rate is fastest and duration shortest (Fig. 1.4b). As temperature rises further above the optimum, rate declines, again more or less linearly, to a maximum (or ceiling) temperature (T_c). Development is more conveniently examined by 1/t than t, because of this linearity.

The relation between 1/t and T between T_b and T_o can be determined by a linear regression of the form

$$1/t = a + bT \qquad (1.1)$$

which is more usefully expressed as

$$1/t = (T - T_b)/\hat{\Theta}_1 \qquad (1.2)$$

where $\hat{\Theta}_1$ is the inverse of the slope b in equation 1.1.

Similarly, the relation between 1/t and T between T_o and T_c can be expressed as

$$1/t = (T_c - T)/\hat{\Theta}_2 \qquad (1.3)$$

The constants $\hat{\Theta}_1$ and $\hat{\Theta}_2$ are here termed *thermal durations* and represent the integral of time and temperature (degree-days) required for the developmental process to be completed[1].

The shape of the response for pearl millet in Figure 1.4b represents that of germination in other species recently examined, with the exception that for chickpea the response is 'flatter' around the optimum. The response also represents later processes, at least between T_b and T_o. In experiments at temperatures above T_o, it is difficult to prevent the air from becoming very dry; and dry air might itself restrict processes independently of temperature.

Analysis of the spread of developmental rate

Since the different percentiles in a sample differ in their developmental rate (1/t), they must differ accordingly in at least one of the attributes – T_b, T_o, T_c, $\hat{\Theta}_1$ and $\hat{\Theta}_2$. The effective variables may be revealed by measuring cumulative development over a range of temperature, and determining T_b, $\hat{\Theta}_1$, etc., as in equations 1.2 and 1.3, for a number of percentiles (e.g. X = 0.1, X = 0.2, etc.). The analysis[1] gives a family of responses, two examples of which are shown in Figure 1.5.

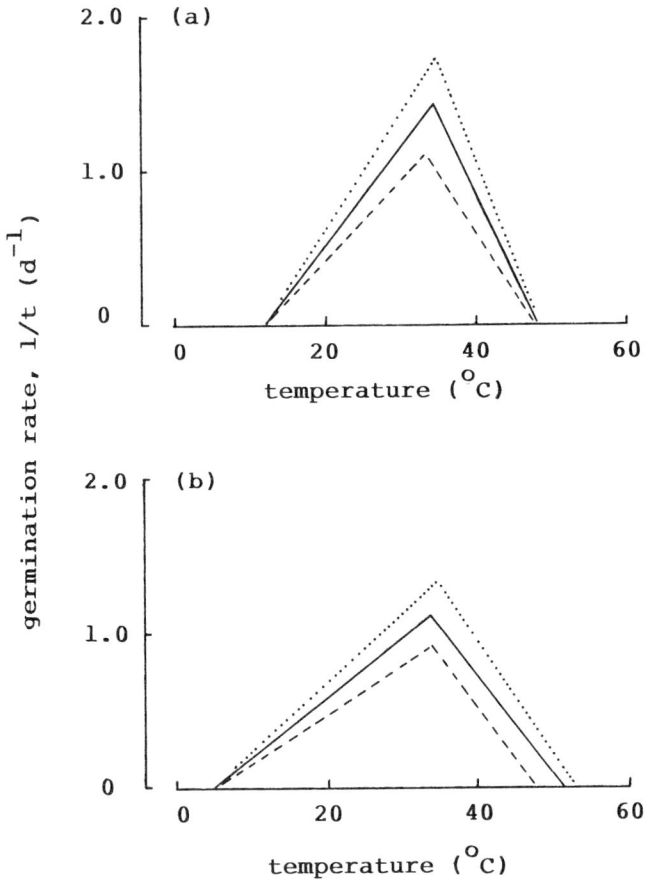

Fig. 1.5. Analysis of the spread of germination rate in a sample of seeds. Effect of temperature on rate (1/t) for the 10th (. . . .), 30th (————) and 60th (– – –) percentile in samples of (a) pearl millet and (b) soybean. Sources: Garcia-Huidobro *et al.*, 1982a, Covell *et al.*, 1986a.

For the species examined – pearl millet and several legumes – the base temperature appeared to be similar for all the percentiles of the population; the differences in rate at any temperature were attributed variously to differences in T_o, T_m, $\hat{\Theta}_1$ and $\hat{\Theta}_2$. For pearl millet (Fig. 1.5a), the thermal durations varied throughout the population whilst T_o and T_m changed little. (Rapidly germinating seeds had a smaller thermal duration and steeper slope than slowly germinating seeds.) The same was found, below T_o, for several legumes, including chickpea, lentil and faba bean, but above T_o the ceiling temperature varied, while Θ_2 was conserved. (Above T_o, rapidly germinating seeds had a higher ceiling temperature than the slowly germinating.) In the legumes, the distribution in the sample of $\hat{\Theta}_1$ below the optimum temperature, and T_c above it, were each approximately normal; and when this is so, the spread in any attribute can be determined from measurements at only a few temperatures[1].

The cellular bases of the differences in the optimum and ceiling temperatures, and the two thermal durations are unknown. Part of the variation in the thermal duration might be a consequence of partial development while on the parent plant or in storage (as distinct from true genetic differences between seeds in the rate of development).

Thermal time

Plants of a genotype grown throughout their lives at different temperatures have a similar base temperature and thermal duration; there seems little adaptation to temperature by these attributes. Even if temperature occasionally falls below T_b, the response is unchanged when temperature returns above T_b. There is limited systematic information on the effects of keeping plants for long periods much below T_b, but tropical crops spend little time in this range.

It is specifically the conservatism of T_b and $\hat{\Theta}_1$ in the face of change in temperature – and throughout a population – that justifies the (usually empirical) practice of expressing the development of a process in terms of a thermal time (Θ) defined by

$$\Theta = \sum_{i=1}^{i=n} (T-T_b) \qquad (1.4)$$

where n is a number of days experienced by the plant at a mean temperature, T (below T_o). The thermal time accumulated during a process has the same numerical value as the thermal duration ($\hat{\Theta}_1$ or $\hat{\Theta}_2$, whichever is appropriate). However, if all the processes of a genotype (e.g. leafing, flowering, podding in a legume) respond to temperature with the same base, optimum and maximum temperatures, then development over a sequence of processes can be expressed in relation to thermal time accumulated from sowing.

The average of the daily maximum and mininum temperatures measured in a meteorological screen is often taken as the mean temperature for the analysis. A correction may be applied to this mean if the temperature during part of a day falls below T_b or rises above T_o; but an alternative relation should be used when temperature is generally above T_o[3]. At temperatures approaching and above T_c, the analysis of thermal time is likely to be very uncertain: the extent of thermal adaptation to these high temperature is unknown for most genotypes.

Figures 1.6 and 1.7 show examples of thermal time analysis, for a population of pearl millet seeds germinating in a laboratory, and a single percentile of maize initiating leaves at three altitudes in western Kenya respectively. In both examples, disparate sets of data were reduced to a single relation, showing temperature alone caused the differences in time to development.

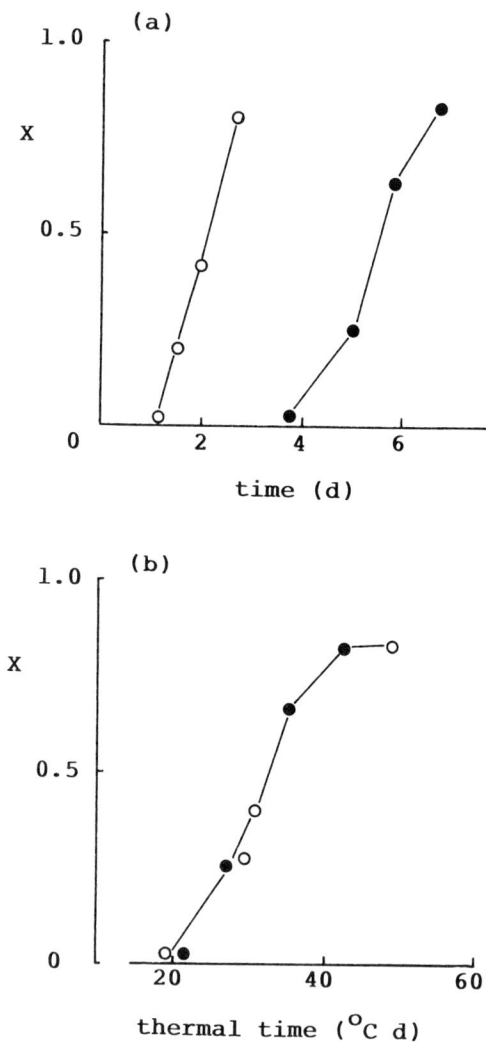

Fig. 1.6. Thermal time analysis: first example.
Cumulative emergence, X, of pearl millet (genotype
'Kala') at temperatures of (●) 16°C and (○) 27°C
expressed in (a) time from sowing, and (b) thermal
time. Source: Mohamed *et al.*, 1988a.

(a)   (b)

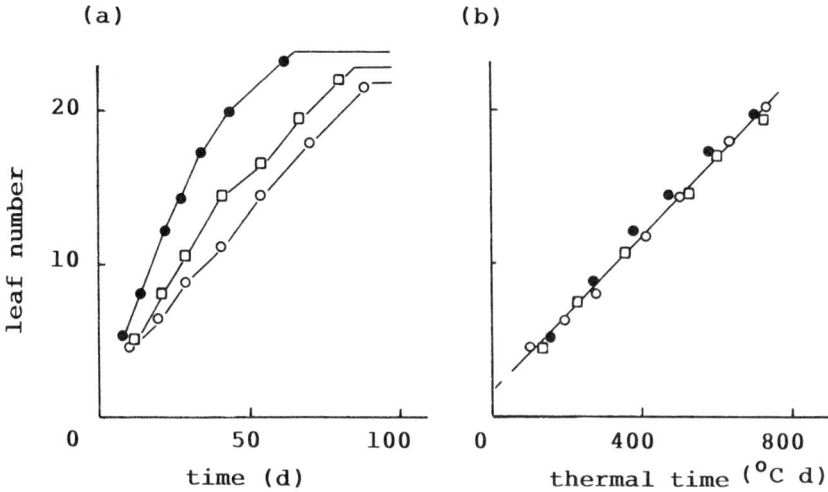

Fig. 1.7. Thermal time analysis: second example. Number of leaves produced on maize culms (mean of population) at three sites in Kenya, at different altitudes (and temperatures): (●) Chemelil, 1268 m, (□) Kitale, 1890 m and (○) Elgon, 2255 m, expressed in (a) time from sowing and (b) thermal time above a base of 12°C. Sources: Cooper and Law, 1978, Cooper, 1979.

More generally, thermal time may be used to detect when other factors are influencing development in an environment when temperature is itself fluctuating.

The analysis of development in relation to thermal time is sometimes used indiscriminately and inappropriately. It requires knowledge of the base temperature, which should ideally be measured in controlled conditions for at least one developmental process. The practice of relating development of tropical species to degree-days accumulated above an arbitrary value, such as 0°C, is inadmissible, as is that of accumulating thermal time over periods when development does not respond to temperature, such as the period of leaf initiation in cereals (p. 23).

The relevance of determinacy

The responses of developmental rate to temperature are intrinsically the same for plants with determinate and indeterminate development. However, determinacy has important implications for the effect of temperature on the duration of a process as a fraction of the period between sowing and harvest. A simple model illustrates the general effects (Fig. 1.8). The durations (as distinct from rates) are shown between sowing and firstly, a developmental event such as the start of flowering, and secondly, harvest.

The time from sowing to the event (identified by the letter 'a' on Figure 1.8) is determinate and governed by temperature. For plants having the *determinate* reproductive habit, the time from sowing to harvest (identified by the letter 'b') also responds as a discrete developmental period since it consists of a series of determinate periods. The times from sowing to the developmental event and to harvest are shortest at the optimum temperature for developmental rate (30–35°C) and lengthen as temperature falls below or rises above this optimum. Accordingly, the *difference* in time between the curves 'a' and 'b' is smallest at the optimum temperature and increases as temperature falls below or rises above the optimum. However, at any temperature, the value on curve 'a' as a fraction of that on 'b' is unaffected by temperature, but equivalent to the ratio of the thermal durations for the developmental process and for the whole life-cycle.

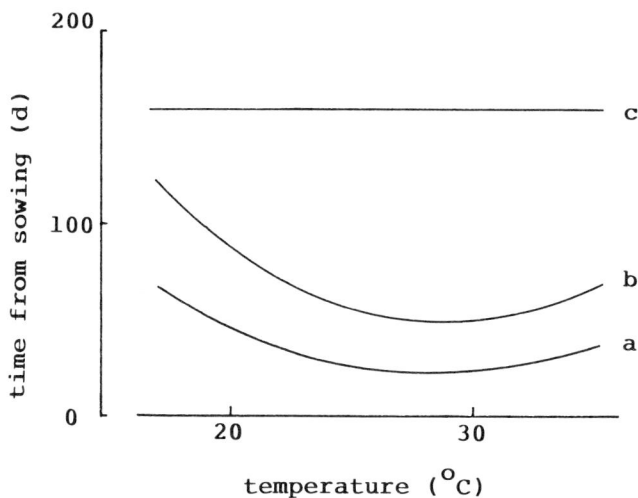

Fig. 1.8. Temperature and duration. Model of the effect of temperature on the duration (t) of the periods between sowing and (a) flowering, and (b) maturity for a determinate genotype, based on attributes of the groundnut genotype Kadiri-3. The line (c) represents the length of the growing season (not temperature dependent).

For plants with an *indeterminate* reproductive habit, the time from sowing to harvest ('c' on Figure 1.8) is not controlled by temperature, as for the determinate, but by other factors – such as a need to harvest at a particular time, or a climatic restriction of the growing season. The time from sowing to the developmental event is governed by temperature as for the determinate (curve 'a'). Therefore, the time interval between the event and harvest 'c' is greatest at the optimum temperature, and decreases as temperature rises above or falls below the optimum – a response opposite in direction to that for the determinate.

These effects of determinacy have very important implications for production and partition of dry matter, as considered in Chapters 2, 3 and 5.

Genetic variation in the response to temperature

The variation has been examined both in the ability of plants to adapt and survive at high temperature (p. 5) and in developmental rate of surviving percentiles. The analysis of rate should take account of variation in the population (p. 7), but is commonly made only for the percentile with the median or the mean rate. Most investigations have been of germination and emergence.

Between species

The information needed to characterize survival and developmental rate, even for germination, has been obtained for a few genotypes of only a few species. Of four examples given in Table 1.1, three, all of which are thought to have originated in the tropics, have similar mean base, optimum and ceiling temperatures. Differences in their developmental rate are caused by differences in the thermal durations. The fourth example, chickpea, has a much lower base temperature (0°C) and higher maximum temperature (> 50°C), but though grown in the tropics is thought to have originated in a mediterranean climate.

The base temperature and the thermal duration, $\hat{\Theta}_1$, are known for many more species, though for only one or two genotypes of each. An extensive investigation of seedling emergence rate[4], showed the base temper-

Table 1.1. Comparison of base (T_b), optimum (T_o) and ceiling (T_c) temperatures for germination of the 50th percentile of the population for different species; means for several genotypes in each sample.

Species	Temperature (°C)			No. of genotypes
	T_b	T_o	T_c	
Sorghum[†]	10	36	47	9
Pearl millet**	11	33	44	8
Groundnut**	10	34	44	14
Chickpea*[‡]	0	30	> 50	5

*Optimum not distinct in some cultivars; measurements not extended above 40°C.
Sources: **Mohamed *et al.*, 1988a. [†]Harris, Hamdi and Terry, 1987. [‡]Ellis, Covell, Roberts and Summerfield, 1986.

ature was between 0–3°C for species originating in temperate climates, and mostly between 9–13°C for those in tropical climates. Considering all species, the base temperature was negatively correlated with the thermal duration, though this correlation was obscured for the tropical species by differences related to seed weight (smaller seeds germinating more rapidly and vice versa). From this and other information, it seems that the large differences in developmental rates (of temperature-sensitive processes) among the important tropical species are attributable largely to differences in the thermal duration. For example, Θ_1 for the period between initiation of successive leaves is about 25°Cd for the rapidly developing pearl millets, 40°Cd for maize, 65°Cd for groundnut, and about 150°Cd for tea.

Within species

Although intra-specific differences in developmental rate have commonly been reported, for processes such as leaf initiation for example, there have been few attempts to define the nature of such differences in terms of T_b, T_o and T_c or the thermal durations. Most of these attempts have considered germination.

In detailed investigations of several genotypes of both chickpea and faba bean[5], there was little intra-specific variation in the base temperature, but much in Θ_1 and the T_o (range 25–35°C), and apparently much in the T_c (which was well above the range of temperature examined in the experiment). However, there were also large differences between genotypes in the spread of attributes within the samples. Among five faba bean cultivars, for example, the thermal times (below T_o) to germination of the 10th and 90th percentiles were about 65 and 100°Cd respectively, a difference of 35°Cd, for the cultivar with the narrowest spread in time-to-germination below T_o, compared to 115 and 220°Cd, a difference of 105°Cd, for the cultivar with the widest spread of time-to-germination.

Less intensive surveys of several other species, including sorghum, pearl millet and groundnut, in which the attributes were measured for the percentile with the median rate, revealed genetic variation in all the attributes, T_b, T_o, T_c and Θ_1, and Θ_2[5]. The range of variation in each of T_b, T_o and T_c was about 5°C. With the additional variation in the thermal durations, germination rate differed between genotypes by up to a factor of two between 25–35°C. There were also indications that the attributes were influenced by the environments from which the cultivars were obtained. For example, the millet genotype with the widest interval between base and ceiling temperature ($T_b = 8°C$, $T_c = 46°C$; an interval of 38°C) came from Niger – a climate of extreme upper and lower diurnal temperatures; whereas the genotype with the narrowest interval ($T_b = 13.5°C$, $T_c = 42°C$; an interval of 28.5°C) came from a coastal area of Senegal, where the diurnal temperature range is much smaller.

On the evidence, the value of an attribute, such as $\hat{\Theta}_1$ or T_c, differs at least as much between the slowest and fastest seeds in a sample from one genotype, as the median value of the same attribute differs among genotypes of the species. Therefore, when defining germination for a genotype, it is just as important to characterize the spread of rates in the population as it is the absolute rate of the median or other percentile.

Correlation between germination and later processes

There is little information to show whether the variation in T_b and the other attributes contributes to the very large differences within species in the timing of later developmental events, including flowering. There is, however, some evidence that the rates of later processes might be linked with the attributes controlling germination. For example, in a comparison of nine sorghum genotypes[5], those classified qualitatively as 'early', germinated most rapidly, and those as late, least rapidly. The mean base temperature was similar for the three categories; differences in rate were attributable to differences in the thermal duration $\hat{\Theta}_1$. Similarly, for eight groundnut genotypes collected from India, Zimbabwe and Brazil, the value of $\hat{\Theta}_1$ for germination was correlated ($r = 0.76$) with the time-to-harvest at the site at which the seeds were collected.

Other investigations have not shown such correlations. For example, among four pearl millet genotypes, differences in the rate of germination were caused variously by differences in the base temperature and the thermal duration. However, germination rates were very highly correlated with rates of emergence and leaf initiation, and with several attributes of expansion and dry matter production[5].

Control by photoperiod (day-length)

The development of only a few species is unaffected by photoperiod. These species are termed day-neutral plants, and include groundnut and some genotypes of cowpea. Within the life-cycle of most species, there is at least one developmental period governed by photoperiod. In some species this period is unaffected by temperature, while in others, it is sometimes affected by it in a more complex way.

The period generally sensitive to photoperiod is that between sowing and the start of reproductive development, but events during that interval sometimes affect the duration of later processes, particularly in cereals. Plants are sensitive to photoperiod during an inductive period, which is preceded and followed by periods in which the plants are not sensitive. Plants are classified according to whether 'long' or 'short' days, experienced during the inductive period, are essential for flowering, or reduce the time taken to flower. Most tropical species are short-day plants, in that the period before flowering is extended if the photoperiod is longer than a certain value (i.e. they require days *shorter* than a certain length if they are to flower in the shortest time). A few tropical species, such as cassava, and many sub-tropical or mediterranean species including chickpea, lentil and barley, that are grown in cool seasons or at altitude in the tropics, are long-day plants: days shorter than a certain value delay or prevent flowering.

Recent research in controlled environments, using mainly tropical and sub-tropical legumes, has defined the response of developmental rate to photoperiod by methods similar to those described earlier for temperature[6]. The general response for short-day plants is shown in Figure 1.9. The developmental rate for (or the rate of progress towards) flowering is maximum (time-to-flowering shortest) and unaffected by photoperiod below a critical value (P_c). The rate decreases – the time-to-flowering increases – as photoperiod rises above that value. In genotypes whose response is termed 'quantitative', there is an upper, or ceiling, photoperiod (P_{ce}) above which further increase in photoperiod causes no further reduction of the developmental rate. In genotypes whose response is 'obligate', the ceiling photoperiod is that above which flowering never occurs.

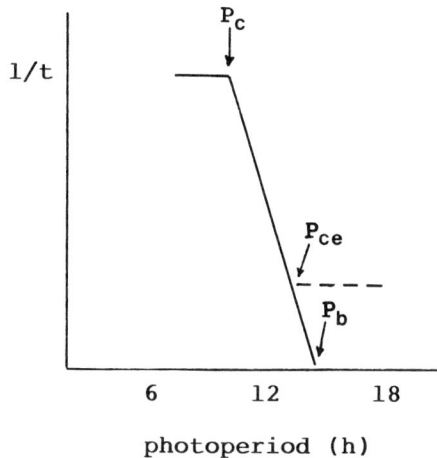

Fig. 1.9. **Idealized response of developmental rate to photoperiod.** Text gives description. From Roberts and Summerfield, 1987 (Further reading, this chapter).

When photoperiod (P) is between the ceiling and the critical values, the relation between 1/t and P can be written in a manner analogous to that for the response to temperature, viz.

$$1/t = (P_b - P)/\varphi \qquad (1.5)$$

where φ has been termed the *photoperiodic time*, or the 'number of hours of light in excess of the base photoperiod which has to be accumulated before flowering occurs'[6], and P_b is the base photoperiod. (P_b and P_{ce} are the same for obligates.)

In this definition of photoperiodic time, the hours of light have to be accumulated during the inductive period. If the limits of this are not known precisely, any hours accumulated before or after the inductive period will be accumulated in error. For many annual legumes, however, the inductive period is relatively long compared with the pre- and post-inductive periods, so the error is small. In cereals, this may not be so.

Genetic variation

Many studies, both in the field and in controlled environments, have shown there is much genetic variation within a species in the control of time-to-flowering. The importance of this control to agriculture is well illustrated in those regions where, to achieve maximum yield, flowering should occur at a particular time in the annual climatic cycle. In Nigeria, for example, the duration of the rains decreases northwards. In each latitudinal zone, the most suitable cultivar (usually the local one) is that which flowers nearest the end of the rains, and fills its grain mostly while growing on moisture stored in the soil. Cultivars that flower before the rains end yield little because the grains turn mouldy; those that flower well after the rains end also yield little but because there is little water left in the soil[7].

The variation in time-to-flowering between cultivars of a species is usually attributed to corresponding variation in the response to photoperiod, though in the field the effects of photoperiod and temperature have rarely been distinguished. Experiments in controlled environments have revealed genetic variation in most attributes of the photoperiodic response: the critical, ceiling and base photoperiods, and the photoperiodic time (φ); and in the phase during development when the plants respond to photoperiod. Recent work has concentrated on the legumes, but much of the work on tropical cereals in the 1960s and 1970s (which did not define genotypes in terms of photoperiodic time, etc.) can now be re-analysed (see p. 22). Among the cereal cultivars, for example, earlies tend to have a very short pre-inductive phase, and a less sensitive response to rise of photoperiod above the critical. Their apex therefore initiates vegetative primordia for a shorter period than that of late genotypes. The degree of earliness is not associated with difference in the rate at which individual vegetative primordia are initiated or expand, so earlies produce fewer leaf primordia than lates[8].

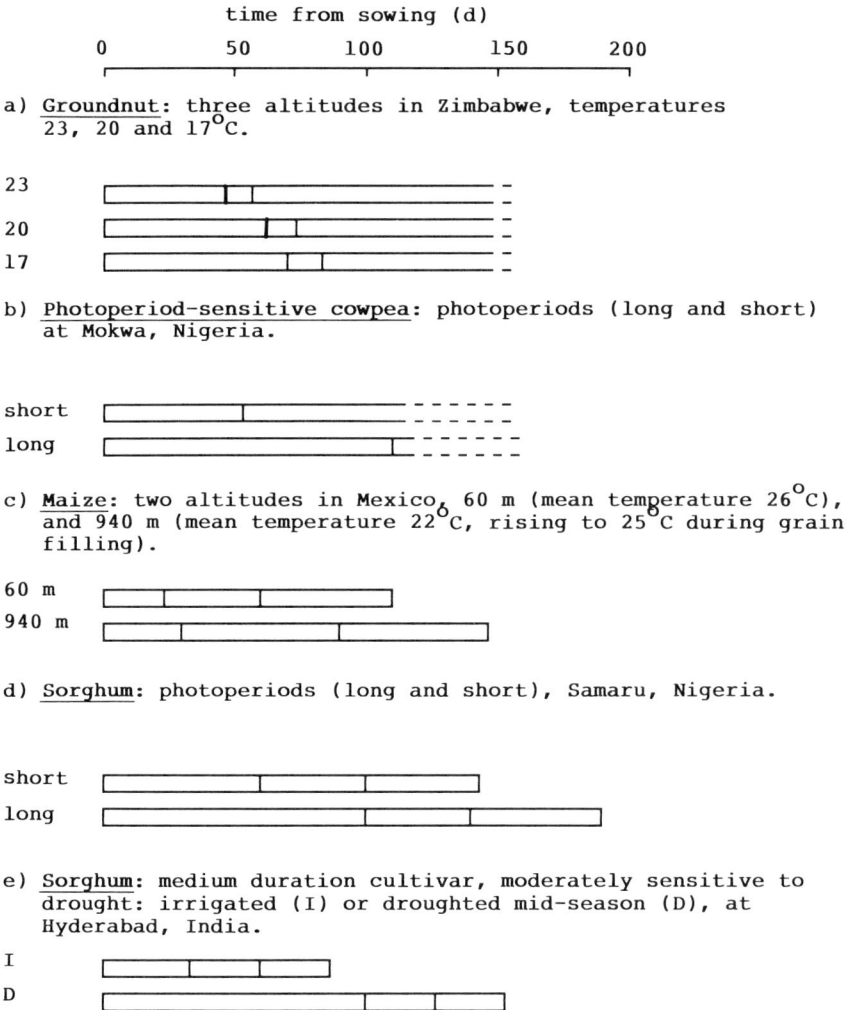

```
                  time from sowing (d)
        0        50       100      150      200
```

a) Groundnut: three altitudes in Zimbabwe, temperatures 23, 20 and 17°C.

b) Photoperiod-sensitive cowpea: photoperiods (long and short) at Mokwa, Nigeria.

c) Maize: two altitudes in Mexico, 60 m (mean temperature 26°C), and 940 m (mean temperature 22°C, rising to 25°C during grain filling).

d) Sorghum: photoperiods (long and short), Samaru, Nigeria.

e) Sorghum: medium duration cultivar, moderately sensitive to drought: irrigated (I) or droughted mid-season (D), at Hyderabad, India.

Fig. 1.10. Examples from the field of effects on development of temperature, photoperiod and drought. The horizontal bars represent the time from sowing to harvest in relation to the scale from 0 to 200 d. For the legumes (a,b), the first segment of each bar indicates the period from sowing to first flower; the second (groundnut only), first flower to first pod. For the cereals (c-e), the first segment indicates sowing to panicle initiation; the second, panicle initiation to anthesis; the third, anthesis to grain maturity. Sources: (a) Williams *et al.*, 1975, (b) Wien and Summerfield, 1980, (c) Fischer and Palmer, 1984, (d) Kassam and Andrews, 1975 and (e) Matthews *et al.*, 1990a.

The work with legumes and some other species[6] has shown the usefulness of the linearity in the response. The attributes of a genotype can be determined from observations of plants grown in only a few different photoperiods.

Examples of the control of development by temperature and photoperiod

The account so far has outlined the principles by which temperature and photoperiod control rates of development. Though both these factors influence most species, temperature has a more general role than photoperiod. The principles are now illustrated by examples from some of the main groups of species. Figure 1.10 shows typical responses in the field.

Indeterminate, sensitive to temperature only

Groundnut and certain genotypes of cowpea are unusual in being insensitive or only very slightly sensitive to photoperiod[9]. Temperature largely governs the timing of development for these plants, and is responsible for seasonal and altitudinal effects such as in Figure 1.10a. The base temperature is similar for the different processes in each species, and for most cultivars is between 8–10°C.

These plants produce leaves for a period of time, both on a main stem and on branches, and then reproductive structures. Generally, leaves continue to be produced during the reproductive period which may continue for several months, so in many environments, both vegetative and reproductive development are, in effect, indeterminate. There seems little quantitative information for cowpea on the control of these processes. In groundnut, the period between the appearance of successive leaves is a developmental duration that responds to temperature as in equation 1.2, with a thermal duration that changes little throughout the life-cycle. The periods from emergence of the seedling to the first branching, and then between subsequent branchings, also respond as in equation 1.2, but little is known of how development along branches is controlled.

Reproductive development is synchronized with leaf development: flowers, pegs and pods are produced after a given number of leaves have been initiated. The groundnut genotype Kadiri 3, for example, flowers when it produces about eight leaves. The thermal duration, Θ_1, for the period from sowing to first flowering by 50% of the population is 720°Cd, which is made up of 80°Cd for emergence and 450°Cd for initiation of eight leaves (for each of which $\Theta_1 = 56°Cd$).

In genotypes whose development is strongly indeterminate, the duration of the reproductive phase responds to temperature as in the general indeterminate case in Figure 1.8. However, for some genotypes, the time from sowing to maturity of most pods, can be treated as a determinate developmental period, governed by a single thermal duration (since it consists of the sum of the thermal durations for initiation of individual leaves on the main stem). Similarly, the period of reproductive development can be represented by another, smaller, thermal duration (consisting of the sum of the thermal durations of the leaves initiated between the start of flowering and maturity of the pods). The duration of the reproductive phase then responds as for the general determinate case in Figure 1.8.

Indeterminate, sensitive to temperature and photoperiod

This category includes most of the important tropical legumes, which are mostly short-day plants, but which exhibit a range of physiological responses to day-length. Effects of temperature and photoperiod have been examined mainly using potted plants in controlled environment chambers. The following account considers two cases of different complexity.

The responses of the photoperiod-sensitive genotypes of cowpea[9] are among the simplest (Fig. 1.11a). When the photoperiod is below a critical value, the timing of the reproductive stages – budding, flowering, podding – is governed by temperature (equation 1.2), as for groundnut and those genotypes of cowpea not sensitive to photoperiod. If the photoperiod rises above the critical value, timing is unaffected by temperature but governed by photoperiod (equation 1.5). The effects of photoperiod can as much as double the time from sowing to flowering in sensitive cultivars (Fig. 1.10b).

The responses of temperature and photoperiod are independent. In any set of conditions, flowering is controlled by the response which gives rise to the slowest developmental rate. In these circumstances, the critical photoperiod, P_c, is that at which the rate controlled by photoperiod is the same as the rate controlled by temperature (i.e. at photoperiods longer than this, the rate is unaffected by temperature). Photoperiod has no influence on the base temperature, but the critical photoperiod is influenced by temperature, being longer at lower temperature. The change of P_c with temperature can be predicted from the equations governing the response of rate to temperature and photoperiod, and differs much between genotypes, from very little to -27 min per °C rise in temperature.

The responses of soybean[9] are more complicated (Fig. 1.11b), and representative of most of the short-day plants. Again the rate of development is controlled only by temperature until the photoperiod rises above a critical level. Unlike for cowpea, the developmental rate above the critical photoperiod is influenced by both photoperiod and temperature. In these cir-

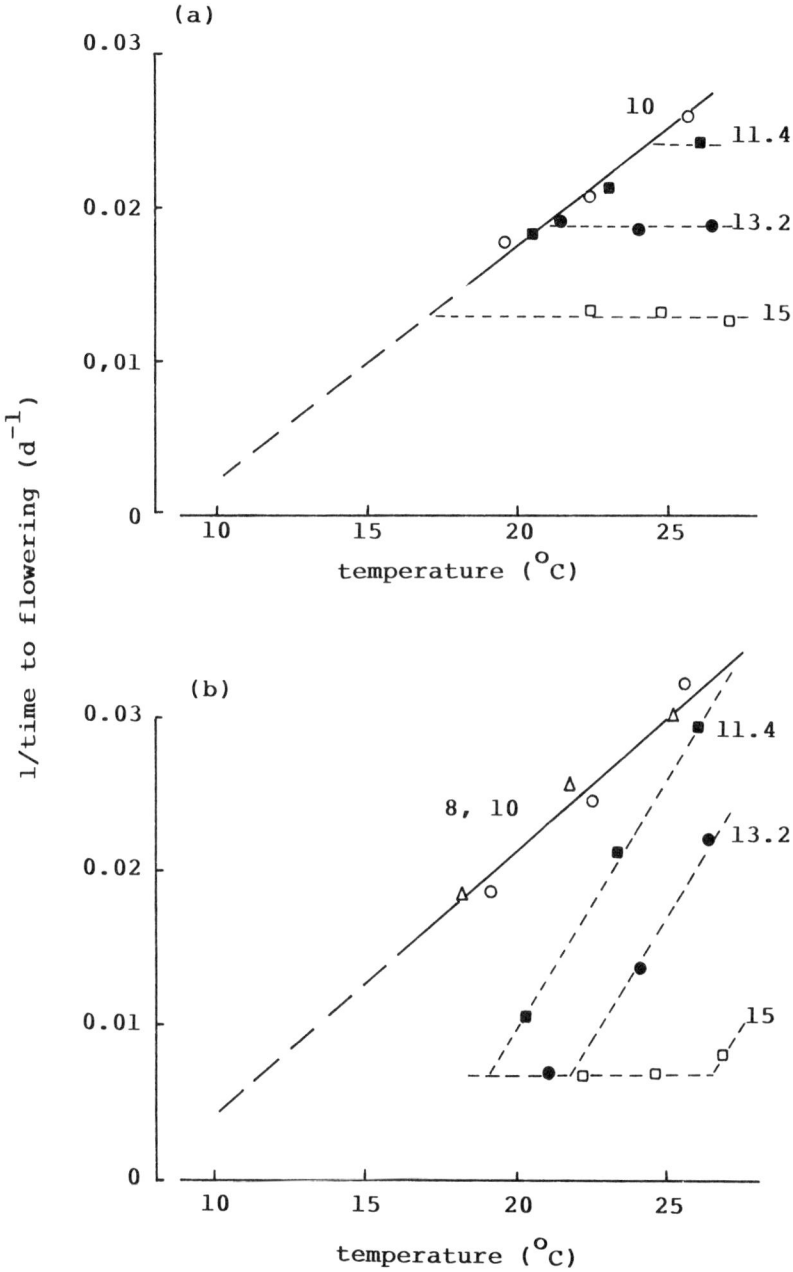

Fig. 1.11. Control of flowering in legumes by temperature and photoperiod.
For (a) the cowpea genotype TVu 1188, and (b) the soybean genotype TGx 46-3c, in
controlled environment chambers. Photoperiods are indicated on the figures.
Sources: Hadley *et al.*, 1983, 1984.

cumstances, the critical photoperiod increases with rise in temperature, while the apparent base temperature increases with rise in photoperiod[6].

Photothermal time

The responses illustrated in Figure 1.11 are two of a range which includes long-day species such as lentil and pea[6]. The complete response for any genotype can be represented as a three-dimensional display (not shown) with developmental rate plotted in relation to temperature and photoperiod. Despite the apparent complexity of these displays, the underlying responses to temperature and photoperiod are linear, so in the region governed by *both* temperature and photoperiod (the 'photothermal response plane'), developmental rate is still governed by a simple equation. Other attributes such as the critical photoperiod and the base photoperiod can similarly be defined by a set of simple equations[6].

Tropical short-day species are grown mostly in combinations of temperature and photoperiod that lie on the photothermal response plane (e.g. responses at photoperiods above 10 h in Fig. 1.11b). Since the relation between developmental rate and temperature remains linear, the period of development can still be expressed as an integral of time and temperature above a base. However, since the base temperature (the value of temperature at any photoperiod when $1/t = 0$) changes with photoperiod, the integral has been termed the photothermal time[6], but has the same units as thermal time (°Cd).

Determinate (cereals), sensitive to temperature and photoperiod

Development of these is commonly considered in three successive stages: that during which vegetative primordia are initiated; that during which reproductive primordia are initiated and expand, and at the end of which flowering occurs; and that in which grains fill and mature. Much of the difference between early and late genotypes lies in the length of the first stage, which ranges from about 17 d in some early pearl millets and sorghums to around 100 d in late cultivars of all the main cereals. The combined period from sowing to flowering ranges from 45 d for short-season types to around 150 d for long-season.

Effects of temperature

With one important exception, a response to temperature of the form in equation 1.2 governs the rate of developmental processes[10]. The response largely causes the slowing of development at higher altitudes and during cool parts of the year (e.g. Fig. 1.10c).

The base temperatures are between 10–12°C for the sensitive processes in most cultivars of the main species. The optimum appears to be about 33°C, but there is little information above the optimum except for germination. The different processes differ mainly in their respective thermal durations (Table 1.2).

Table 1.2. Development in a short-season pearl millet hybrid (BK 560): thermal durations (Θ_1) for 50% of the population. The base temperature was about 10°C for all processes.

Development	Thermal duration in degree-days Θ_1(°Cd)
Between successive leaves and root axes	26
Sowing to first tiller	200
Between successive tillers	80
Expansion of tiller leaves	270
Increase in mass of tiller leaves	420
Between successive spikelets	0.23
For all spikelets	190
Floral initiation to anthesis	460*
Expansion and growth of main stem	360*
Increase in weight of panicle	550
Increase in weight of grain	290

*Process may be indirectly affected by photoperiod.
Sources: Ong, 1983a, b; Squire, 1989a, b.

The exception is the period – immediately after seedlings have emerged – in which vegetative primordia are initiated. Temperature still has some control over this phase and, in a few cultivars, controls it much as it controls other phases. But generally, temperature has little effect over a broad range between 20–30°C. In this range, there seems to be a restriction imposed on developmental rate to prevent the time from sowing to the start of reproductive development from becoming very small at temperatures around the optimum (Fig. 1.12). Moreover, this response in most cultivars is modified by photoperiod.

Effects of photoperiod

The photoperiod influences the length of the phase of leaf initiation just referred to, which ends when a panicle is initiated[11], but has little direct effect on later phases. In photoperiod-sensitive cultivars, the effect can be very large. In Nigeria, for example (Fig. 1.10d), the time from sowing to panicle initiation of a long-season sorghum decreased from 98 to 58 d as stands emerged into increasingly shorter photoperiods[12].

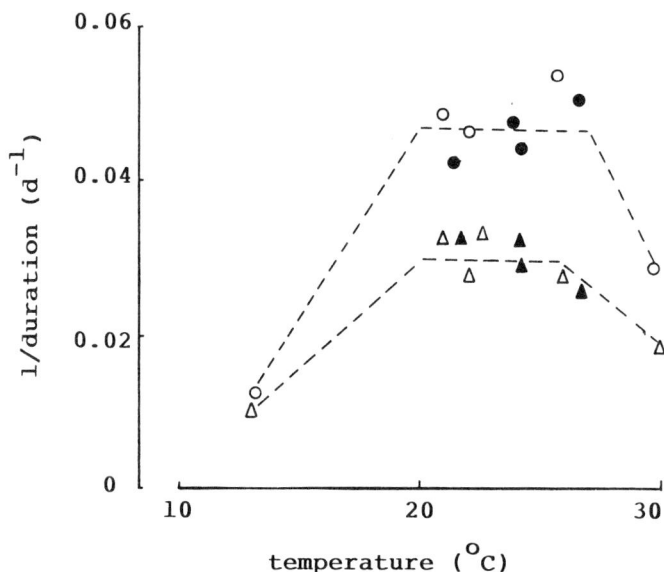

Fig. 1.12. Example of the control of flowering in cereals by temperature and photoperiod. For the sorghum genotype Early Hegari. The period between sowing and initiation of floral primordia is shown as a rate (1/duration) at various combinations of temperature and photoperiod (circles, 10 h, triangles, > 14 h). The dashed lines indicate an interpretation of the responses. Sources: open symbols, Table 2 in Quinby, Hesketh and Voigt, 1973; closed symbols, Figures 2, 4 and 5 in Caddell and Weibel, 1971.

The pre-inductive phase, during which photoperiod has no influence on the time to the start of panicle initiation, can be as short as two or three days after sowing in early genotypes (i.e. until about three leaves have been produced), or more than 40 d in late genotypes. Once the plants are sensitive to photoperiod, the panicle is initiated in the shortest time when day-length is shorter than a critical value of about 12 h (though the critical photoperiod differs between cultivars). Day-lengths longer than the critical delay the start of panicle initiation, but do not prevent it in many genotypes, which are therefore quantitative short-day. The minimum inductive period is about 14 d when the photoperiod is 12 h.

The control by photoperiod of the time to panicle initiation has rarely been systematically analysed in tropical cereals, as the time-to-first-flower has in the legumes. Nevertheless, the reported experiments in the literature provide a rich source of material for analysis (e.g. Fig. 1.12). The way photoperiod modifies the response to temperature differs greatly among cultivars[13]. Most commonly, between temperatures of 20–30°C, photoperiod governs the duration as in the idealized response in Figure 1.9. When 1/t is expressed in relation to temperature, the resulting family of responses has similarities to that for cowpea in Figure 1.11a, in that photoperiods increasingly longer than the critical impose a progressively lower ceiling on the rate (i.e. ensure a longer minimum time before panicle initiation). The curves differ from that for cowpea in that longer photoperiods also increase the thermal duration (i.e. reduce the slope), though appear not to affect the base temperature.

In some, generally long-season types, that develop slowly, photoperiod does not impose a ceiling on the rate as in Figure 1.12, but influences only the thermal duration (which increases as photoperiod increases). Whatever the form of response, long photoperiods act as if to prevent the period from sowing to flowering being very short when temperature is near the optimum.

The post-inductive period

When the panicle is initiated, the apical meristem and the plant as a whole are still small. During the next few weeks, the panicle differentiates and enlarges, and is then raised as the stem internodes elongate. Flowering occurs as the panicle emerges from the uppermost leaf sheath. The period from panicle initiation to flowering is unaffected by the current photoperiod, and is therefore a post-inductive period which, for short-season cultivars, is generally longer than the inductive period itself.

The temperature and photoperiod experienced by the plants during the inductive period influence the number of leaves initiated during this period. The developmental rate for the formation of leaf primordia is governed by temperature according to equation 1.2, and is not much affected by photoperiod. So at a given photoperiod, more primordia are initiated before the end of the inductive period as temperature rises from T_b to T_o; and at a given temperature, more are initiated as photoperiod rises above the critical (see p. 44 for further analysis).

The overall control of the period between sowing and flowering can be treated as a variation of the general determinate case in Figure 1.8. Long photoperiods increase the time from sowing to flowering (curve 'a' on Figure 1.8), but have little effect on the period between flowering and maturity (the difference between curves 'a' and 'b'). High temperatures (between T_b and T_o) before flowering have two opposing effects: more leaves are initiated up to the end of the inductive phase, but each lamina and stem internode expands for a shorter period of time. The latter response seems generally the stronger in the field, such that the period between sowing and flowering shortens as temperature rises[10].

Indeterminate, perennial species

The developmental responses of some perennial vegetative and fruit crops have been examined in detail, but seldom with the analyses used in this chapter. Developmental periods of cassava, a storage root crop, are controlled by temperature to some extent as such periods are in the cereals and legumes. The base temperature is around 10°C, though the time-to-first-flowering and branching appears to have a low optimum around 24°C[14]. Cassava is a long day-plant: both flowering and first branching occur earlier in long days[14]. These effects of temperature and photoperiod on development of aerial vegetative structures have important implications for partition of assimilate to the storage roots, as considered in Chapter 5.

Development in the tropical fruit and beverage crops usually proceeds intermittently (see Further reading, this chapter). In some species, such as cocoa, leaf development might alternate with root development, or with reproductive development. In others, such as oil palm, leaves and inflorescences are produced continuously, except during drought, though there are periods in which most of the new inflorescences are either all male or all female. Such intermittent development can sometimes be traced to seasonal changes in climate, but it still occurs in some species when there is very little seasonal change. There is evidence that it might be related to cycling of the store of assimilate or nutrients in the plant and, further, that a period of untypical climate might initiate a long period of cyclical development.

The implied role of assimilate storage in the control of development in perennials means that work in controlled environments with cuttings or potted plants might be largely irrelevant. Nevertheless, there is no reason, in principle, why development of woody plants in the field should not be examined by relations between 1/t and temperature or photoperiod. This approach has been used with tea in Malawi and Kenya to investigate seasonal and geographical variation in shoot production[15]. The period required by a small bud to grow to a shoot of harvestable size was treated as a typical developmental duration. The rate of development (1/t) was little affected by altering the day-length by means of artificial lighting strung over a crop, but was strongly governed by temperature, as in equation 1.2, with a base of 12–13°C and a thermal duration of 450–500°Cd. The seasonal and altitudinal variations in yield were caused mainly by effects of environment and management on developmental rate, rather than by an intrinsically intermittent pattern of shoot growth. Similarly, the duration of pod filling in cocoa has been defined by a base temperature around 9°C and a thermal duration of 2500°Cd[15].

Restriction by other factors

Information on the sensitivity of survival and developmental rate to other factors, chiefly solar radiation, drought and shortage of nutrients, is largely qualitative. The systematic responses have not yet been revealed as they have for temperature and photoperiod.

Solar radiation and assimilate supply

The variations with season and altitude in the incoming solar radiation have very little effect on the timing of a developmental sequence. Even the shaded partners in mixed crops, perhaps receiving one-half or one-third of the incoming radiation, usually develop at a similar rate to corresponding unshaded crops (References in Chapter 2). Development seems to be affected only by covering crops with a dense shade.

The importance of assimilate supply, as distinct from solar radiation itself, can usually be revealed by observing the response to change in the plant population density. The plant population determines the fraction of the dry matter produced that is available to each individual. As population increases, and this fraction becomes smaller, some of the individuals (or a part of most individuals) cease to develop; or development ceases for some processes but not others (e.g. fruits but not leaves).

In some species, photoperiod can modify these effects of population density. Photoperiods that increase the duration of vegetative development on the main axis of a plant tend to reduce survival of, or delay development on, axillary structures (tillers, branches) – an effect probably operating through the rate of assimilate supply to these structures[16].

The extent to which stands regulate the number of developing individuals or structures to match the assimilate supply depends largely on their gross morphological attributes[17]. Pigeon pea, for example, is able to reduce the number of growing-points and thereby the mean mass of individuals, so even at very high populations, few die. Many cereals, and especially rice, compensate by regulating the number of tillers. One of the main effects of this compensation is that the timing of development on the remaining stems is conserved over a wide range of assimilate supply. However, the palms, with their rigidly symmetrical architecture, are unable to compensate by reducing the number of axes (since there is only one). Even so, their leaf production rates change little over a wide range of population density: compensation is achieved by an increase in the fraction of inflorescences that are male (and so consume much less dry matter than fruit bunches).

Drought and nutrient shortage

Dry, infertile soils could affect development either directly, by restricting cellular processes in the apices, or more or less indirectly, by restricting the production of new assimilate. Generally, these effects are not distinguished.

Many plants respond to these adverse conditions by reducing the number of developing apices, thereby conserving processes on the main apex (cf. response to assimilate shortage). Generally, development on this apex is far less sensitive than expansion and dry matter production.

The thermal responses described earlier can be used to define a response to drought or nutrients, or to compare responses at different temperatures. This approach has seldom been taken[18]. Even when is has, the assumption has been made that the base temperature is unaffected and only the thermal duration changes (though this is a reasonable assumption).

Effects of drought

There have been many attempts to express developmental rate (and other rates) in terms of the leaf water potential Ψ_1, solute potential Ψ_s (both negative), or turgor potential Ψ_p (positive: the difference between Ψ_s and Ψ_1). This approach has sometimes shown rate to decrease more or less linearly with fall in the turgor potential (and sometimes water potential) above a base potential, Ψ_{pb}. By analogy with the responses to temperature and photoperiod, developmental rate should then be related to turgor potential by

$$1/t = (\Psi_p - \Psi_{pb})/\Phi \qquad (1.6)$$

where Φ is the accumulated integral of time and turgor potential above the base potential required for the process to be completed.

Relations of this form have been found or implied[18], but are much less common than the corresponding relations for temperature and photoperiod. The attributes Φ and Ψ_{pb} are generally not conservative, probably for several reasons. (See Chapters 2 and 3 for discussion and examples for expansion and growth processes, and 'Further reading' for these chapters for recent accounts of leaf water relations.)

There have been few attempts to define genotypes in terms of Φ or Ψ_{pb}, or comparable attributes. However, a wide range of qualitative responses exists, even within a species. For certain genotypes, the form of response depends on the severity of the drought, which is also usually defined qualitatively. Development of cowpea, for example, may be unaffected, or even accelerated, by mild or moderate drought, but slowed or suspended by severe drought[19]. Among the cereals[20], some genotypes, particularly short-season pearl millets, are very little affected by dry soils and atmospheres: their main culm develops at much the same rate irrespective of the degree of dryness, and even when dry matter production is negligible. Development in many other genotypes is delayed by drought (e.g. rice), and in others still might be suspended for several weeks or months (e.g. Fig. 1.10e). In the latter category are the long-season sorghums whose development may cease even before the panicle is initiated, and groundnut, whose pegs will accumulate on a hard soil surface.

For genotypes whose development is suspended during a drought, it matters whether development will resume if there is later rain or irrigation. In this respect, genotypes again differ much – even types that develop similarly in moist conditions. Those that survive, and continue developing, commonly have distinguishing physiological attributes, among which are asynchronous reproductive development among the different tillers or branches, the ability of leaves to avoid solar radiation, and the resistance or tolerance of tissue to dehydration (Chapter 2).

Some genotypes respond so consistently to drought that, to a degree, they can be described as 'drought-avoiding' or 'drought-tolerant'[21]. Both types suit particular sets of environmental conditions: the avoiders are superior in the short rainy season or on a limited store of moisture; the tolerators suit those climates with a defined mid-season drought. Among the most versatile are those cowpeas that have both avoiding and tolerating attributes – developing more rapidly if the drought is mild, but suspending development during a severe drought, and resuming it, as if unaffected, if the drought is relieved[19].

Effects of nutrient shortage

There is little information on how development is governed by the concentration of nutrients in plants (Chapter 2 gives further discussion). However, nutrient shortage in the soil has consistent effects in experiments with cereals: the period between sowing and flowering increases and that of grain filling changes little or decreases. In all experiments, nutrient shortage reduces the fraction of the life-cycle spent in the reproductive phase.

The lengthening of the period before flowering is associated with an increase in the time between the appearance of successive leaves. The increase was large for pearl millet grown in pots[18]: the thermal time for the appearance of a leaf increased from 60°Cd in the treatment with the most nutrients to 97°Cd in that with the least. The effects reported in the field for barley, maize and sorghum are small or moderate[22]. For example, withholding fertilizer from barley in Syria (which as much as halved dry matter production) had no effect on the time to anthesis at one site, but increased the time by 13 d from 139 d to 152 d at another site. The shortening of grain filling seems less common. It did not occur in the stands of barley and sorghum. In the maize, the period of grain filling in unfertilized stands was about 60% of that in stands given most fertilizer. (The physiological cause and the implication of this shorter duration are examined in Chapter 5.)

Concluding remarks

There is now a set of principles for examining the responses of developmental rate to temperature and photoperiod. The basis of these principles is that rate, expressed as 1/t, is generally related *linearly* to temperature or photoperiod over a wide range. This linearity means that:

- development can be expressed in terms of an integral such as thermal time, photoperiodic time or photothermal time – a form of analysis that is very valuable for modelling development and assessing the extent to which development is affected by other factors in the field;
- the developmental responses of a genotype can be described in terms of a small number of attributes (T_b, etc.);
- the responses can be defined from measurements in only a few combinations of temperature and photoperiod[1,6,9].

The principles have been developed mainly in controlled environments. There is accumulating evidence that the same principles govern development in the field, but corroboration is required for more species in more environments, particularly for photothermal responses and at temperatures above the optimum, where models developed largely between the base and optimum temperatures might be invalid[23]. For many species, the only evidence for responses above the optimum is for germination, and it has to be assumed that developmental rates of later processes respond in this range with optimum and ceiling temperature the same as for germination.

The principles would provide a sound basis for defining the responses of development to restrictive factors. So far, there has been little use of them in this way. From other, mainly qualitative analyses, it seems development is hardly affected by variations in solar radiation, slightly or moderately affected by nutrient shortage, and sometimes greatly affected by drought. Generally, however, these restrictive factors are less important than temperature and photoperiod in governing development of tropical crops.

References

1 Garcia-Huidobro *et al.*, 1982a, b; Covell *et al.*, 1986, Ellis, Covell, Roberts and Summerfield, 1986, Ellis, Simon and Covell, 1987.
2 Susceptibility during imbibition; Garcia-Huidobro *et al.*, 1985. Protein synthesis; Ougham and Stoddart, 1985, Ougham, Peacock, Stoddart and Soman, 1988. Thermal tolerance of sorghum, with references to other species; Ougham and Stoddart, 1986.
3 Correction for low temperature, with pocket calculator; Snyder, 1985; with computer (example for tea); Tanton, 1982a. For above the optimum; Garcia-Huidobro *et al.*, 1982b.
4 Angus *et al.*, 1981.

5   For chickpea and faba bean; refs by Ellis *et al.*, in [1]. For pearl millet and groundnut; Mohamed, Clark and Ong, 1988a. For sorghum; Harris, Hamdi and Terry, 1987.

6   Hadley *et al.*, 1983, 1984. For a general summary of photoperiodic effects, the review by Roberts and Summerfield, 1987. (See Further reading, this chapter.)

7   For sorghum; Curtis, 1968; Andrews, 1973. For legumes; Wien and Summerfield, 1980.

8   For time-to-flowering and leaf number among genotypes; Evans, Visperas and Vergara, 1984 (rice). Algarswamy and Bidinger, 1985 (pearl millet). See also p. 45.

9   Groundnut; Leong and Ong, 1983, Harris *et al.*, 1988. Cowpea; Hadley *et al.*, 1983. Soya bean; Hadley *et al.*, 1984. The papers by Hadley *et al.* give the equations for determining the ceiling and base photoperiods, the photoperiodic time, and other attributes of the response to photoperiod.

10   Ong, 1983a, b, gives the analysis for pearl millet, and references to maize and sorghum, but see also Coaldrake and Pearson, 1986. For altitudinal effects, maize; Cooper, 1979, Cooper and Law, 1978, Fischer and Palmer, 1984. Altitude, rice; Chamberlin and Songchao Insomphun, 1982.

11   For pearl millet; Begg and Burton, 1971, Carberry and Campbell, 1985, Craufurd and Bidinger, 1988. Sorghum; Caddell and Weibel, 1971, Quinby, Hesketh and Voigt, 1973. Maize; Coligado and Brown, 1975. Rice; Vergara and Chang, 1976, also ref [8]. See also Hesketh, Baker and Duncan, 1972.

12   Kassam and Andrews, 1975.

13   For an illuminating illustration of genetic variation in the *form* of the response to temperature at long and short photoperiods, try analysing, in terms of 1/t, the times to floral initiation of maize given in Table 2 of the paper by Quinby, Hesketh and Voigt, 1973.

14   Cassava; Keating and Evenson, 1979 (temperature), Irikura *et al.*, 1979 (altitude; the response to temperature of early leaf initiation in their Figure 4 seems linear with an extrapolated base of about 10°C), Veltkamp, 1985 (photoperiod).

15   Tea; Squire, 1979 (temperature), Tanton, 1982a, b (temperature and photoperiod). Cocoa; Alvim, 1977.

16   Carberry *et al.*, 1985, Carberry and Campbell, 1985.

17   Examples for species mentioned in this paragraph. Pigeon pea; Rowden *et al.*, 1981. Cereals; Ong, 1983a. Cassava, Fukai *et al.*, 1984. Oil palm; Corley, 1973; Breure, 1988a, b.

18   Examples of the thermal time analysis; Harris *et al.*, 1988 (groundnut, drought/water potential), Coaldrake and Pearson, 1985a (potted millet/nutrients).

19   Grantz and Hall, 1982.

20   Cereal genotypes; Puckridge and O'Toole, 1981 (rice), Mahalakshmi and Bidinger, 1985a, b, 1986 (pearl millet), Rees, 1986a (sorghum, Botswana), Matthews *et al.*, 1990a, b (sorghum, India).

21   Hall *et al.*, 1979.

22   Gregory, Shepherd and Cooper, 1984 (barley, Syria), Lemcoff and Loomis, 1986 (maize), Muchow 1988a, b (maize, sorghum).

23   For example Dow El-Madina and Hall, 1986.

Further reading

The review by Roberts and Summerfield gives a comprehensive and detailed account of recent analyses of the responses to temperature and photoperiod, mainly for herbaceous crops. A companion chapter in the same volume concentrates on the physiology of the photoperiodic response. Both these chapters contain references to the standard works on temperature and photoperiod. For tree crops, Browning considers the hypotheses for control of vegetative and reproductive growth. Information on individual tree species, such as cocoa, coffee, tea and oil palm, can be found in the books edited by Alvim and Kozlowksi, and Sethuraj and Rhagavendra. The *Handbook of Flowering* contains background information on the physiology of flowering in individual species.

Alvim, P. de T. and Kozlowski, T. T. (eds) (1977) *Ecophysiology of Tropical Crops.* Academic Press, New York.

Browning, G. (1985) Reproductive behaviour of fruit tree crops and its implications for the manipulation of fruit set. In: Cannell, M. G. R. and Jackson, J. E. (eds) *Trees as Crop Plants.* Institute of Terrestrial Ecology, Monks Wood Experimental Station, Huntingdon, PE17 2LS, UK, pp. 409–25.

Halevy, A. H. (ed.) *Handbook of Flowering* , CRC Press Inc., Boca Raton, Florida, USA.

Roberts, E. H. and Summerfield, R. J. (1987) Measurement and prediction of flowering in annual crops. In: Atherton, J. G. (ed.) *Manipulation of Flowering.* Butterworths, London, pp. 17–50.

Sethuraj, M. R. and Rhagavendra, A. S. (eds) (1987) *Tree Crop Physiology.* Elsevier, Amsterdam.

Summerfield, R. J. and Roberts, E. H. (1987) Effects of illuminance on flowering in long- and short-day grain legumes: a reappraisal and unifying model. In: Atherton, J. G. (ed.) *Manipulation of Flowering.* Butterworths, London, pp. 203–23.

Chapter Two

The Leaf Canopy and Root System

Introduction

This chapter considers the effects of environmental factors, principally temperature, water and nutrients, on the size and longevity of the shoot and root systems. Both systems are composed of many sub-units (e.g. leaf laminae, root branches), but are conventionally described in terms of an area of foliage and a length of root respectively. The complete shoot system is commonly defined by a leaf area index, L, given by

$$L = N_p n_s a_s \qquad (2.1)$$

where N_p is the number of plants per unit ground area, n_s the number of leaves and other foliar units per plant and a_s their mean area. The complete root system is expressed here as a root length per unit area of ground (R) given by

$$R = N_p n_r l_r \qquad (2.2)$$

where n_r is the number of root units (e.g. secondary, tertiary branches) per plant, and l_r is their mean length.

The principal sub-units of most crop canopies are the leaf laminae, but leaf sheaths and panicles also contribute much to the canopies of cereals. The different sub-units of the root system are less easily distinguished than those of the canopy to an extent that second, third and higher order lateral roots are usually grouped in estimates of R. (The root systems of species in mixed crops are also very hard to distinguish.) The analysis of root systems also takes into account the depth they reach, and their proliferation in different soil layers above that depth. This proliferation can be expressed initially in terms of the distance from the row reached by the leading roots and, later when roots of adjacent rows mingle, as a length of root per unit soil volume (l_v).

Expansion, stability and senescence

The change of leaf area and root length during the life of a stand is represented in Figure 2.1. After a period of slow initial increase (phase I on the figure), L (or R) rises rapidly and more or less constantly in time (phase II) to a maximum (phase III). The relation between L and time during phases I and II may be described by a logistic curve[1], or simplified by assuming a more or less linear relation between size and time during much of phase II. After a further period of time, L declines from the maximum (period IV), as old tissue is lost more rapidly than new is produced. For annuals, the maximum size of the systems is usually achieved when expansion has ceased and before much tissue has been lost. For perennials, the maximum is the balance between the production for new, and loss of old, tissue.

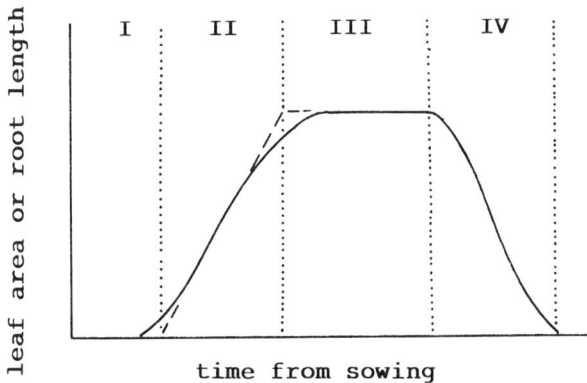

Fig. 2.1. Representation of change in time of leaf area index or total root length. Text gives description.

The proportions of the life-cycle spent by plants in each of the phases I to IV depends on growth habit and environmental conditions. The length of phase III is particularly variable. It is usually short for annuals, for which loss of leaf begins, and sometimes overtakes gain, during expansion. If provided with ample water and nutrients, some indeterminate annuals continue expanding until they are harvested, so do not enter phase III. In contrast, perennials spend much of their life in phase III, and are generally uprooted before they enter phase IV. The cycle of expansion, stability and senescence might be repeated if stands lose some leaf area or root length during a drought.

The size and longevity of leaf canopies and root systems are now examined in terms of the rates and durations of the processes controlling the phases in Figure 2.1. First the canopy is examined, then the root system, and finally the interdependence of leaf and root. To emphasize this interdependence, the account makes much use of a series of experiments with pearl millet and groundnut, in which both L and R were measured for stands in different soils and climates. As in Chapter 1, responses to temperature are the basis for examining the effects of other factors.

The leaf canopy: the importance of temperature

Many of the effects of temperature described in Chapter 1 influence leaf area index. Effects on the survival of individuals probably dominate all other effects when the temperature is below 15°C or above 40°C. Between these limits, temperature governs the durations of the many processes that determine the sequence of phases in Figure 2.1. The most investigated of these processes has been the time to emergence of a plumule or a shoot arising from a sett (e.g. cassava)[2]. This time depends much on sowing depth as deep sowing increases the thermal duration for a percentile of the population and reduces the fraction of the population that emerges (Fig. 2.2). Moreover, the increase in thermal duration is greater for the more slowly developing individuals, so the spread of thermal time between the first and last seeds to emerge increases.

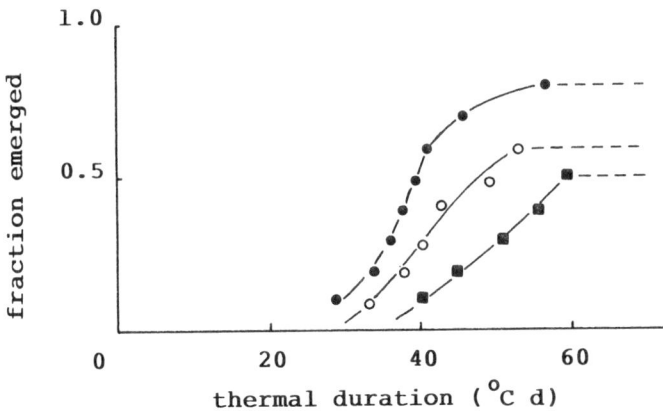

Fig. 2.2. **Emergence and soil depth.** Thermal duration $\hat{\Theta}_1$ for emergence of different percentiles of the population (X = 0.1, etc., see Fig. 1.3, equation 1.2), for pearl millet seeds sown at depths of: •, 20 mm; ○, 40 mm; ■, 60 mm. Source: Reference 2.

For seeds sown between 2–3 cm below the soil surface, the thermal duration, $\hat{\Theta}_1$, for the 50th percentile ranges from 40–80°Cd among the main tropical crops which is equivalent to roughly only 5% of the combined length of phases I and II[2]. For the surviving individuals, later processes are considerably slower than emergence and much less is known about them.

The effects of temperature on the later processes that control the size and longevity of the canopy are examined in the next section. The individual elements are considered first, followed by the system as a whole.

Expansion and size

Temperature strongly affects the rate of expansion of most leaves and stems[3]. When expressed as a change in the length, or area, of tissue per unit time ($\delta l/\delta t$), the rate usually increases linearly with temperature (T) above a base temperature (T_{br}) as in

$$\delta l/\delta t = \rho_1(T - T_{br}) \tag{2.3}$$

where ρ_1 is the *thermal rate* of extension or expansion (units such as mm $(°Cd)^{-1})$[3]. The equation is valid up to an optimum temperature at which rate is fastest. Measurements on plumules and radicles in controlled environments suggest that, above the optimum, the rate decreases linearly with rise in temperature, reaching zero at a ceiling temperature. The shape of the response is therefore similar to that for l/t shown in Figure 1.4b. However, there is little reliable information for temperatures above the optimum for the responses of larger structures.

The size of a structure is the product of the mean rate and the duration of expansion, and there are two main types of response. In determinate structures (Fig. 2.3a), the duration responds to temperature as in equation 1.2 (Chapter 1), and the size (Y_f), when expansion has ceased, is given by

$$Y_f = \rho_1\Theta_1(T-T_{br})/(T-T_{bd}) \tag{2.4}$$

where Θ_1 is the thermal duration for the process of expansion and T_{bd} the corresponding base temperature. If the base temperatures for rate and duration are the same, the effects cancel, and size becomes independent of temperature. In indeterminate structures (Fig. 2.3b), the duration of expansion is unaffected by temperature, and size increases as temperature rises, as in

$$Y_f = \rho_1 t_f(T-T_{br}). \tag{2.5}$$

At harvest, structures are therefore largest at the optimum temperature for expansion rate.

Size of individual elements

The logistic curve[1] is commonly used to describe the change with time in the length or area of a leaf, but during much of the extension time, the rate can be treated as constant (in a unchanging environment). The rate can be meas-

(a)

Y_f (warm and cool)

size

warm

cool

time

(b)

Y_f (warm)

size

Y_f (cool)

warm

cool

t_f

time

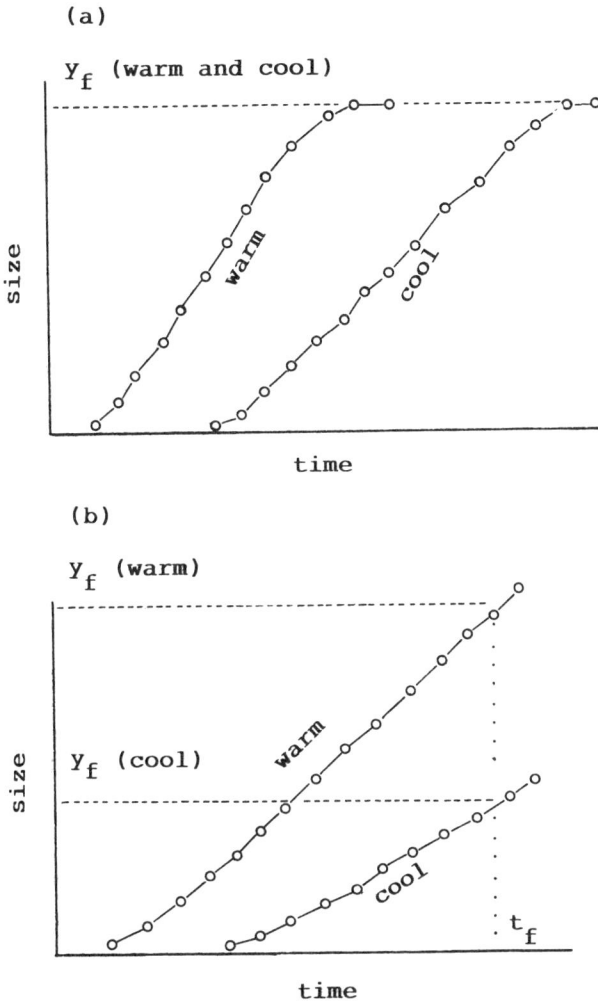

Fig. 2.3. Representation of the effect of temperature on the size of structures with (a) determinate and (b) indeterminate habit. For the determinate, effects on rate and duration compensate; the final size of the structure (Y_f) is unaffected by temperature. For the indeterminate, the duration is unaffected by temperature; at any chronological time (t_f) after expansion begins, Y_f is larger in the warmer environment.

ured over several days, using a ruler for example, or, in species with rapidly extending leaves, over 30–60 min with electronic auxanometers. Rate is usually strongly governed by temperature as in equation 2.3 (e.g. Fig. 2.4a). For the majority of the tropical species that have been examined, the base

(a)

(b)

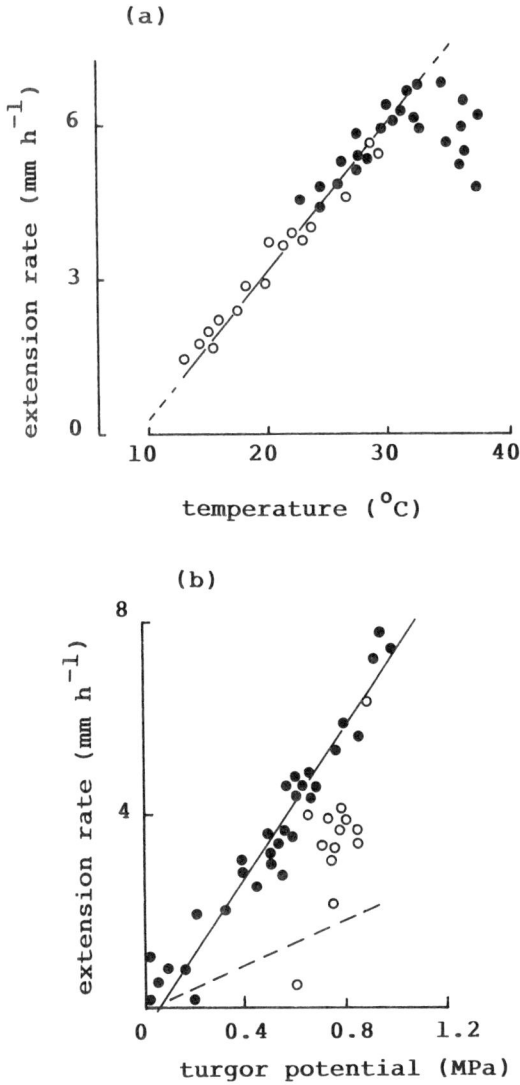

Fig. 2.4. Control of leaf extension rate. For pearl millet grown in field soil within controlled environment glasshouses: (a) extension rate and meristem temperature at mean air temperatures of ($\circ$) 19°C and ($\bullet$) 31°C, (b) extension rate and leaf turgor potential, at near optimum temperatures, when solar irradiance was > 100 W m^{-2} ($\bullet$) and < 100 W m^{-2} ($\circ$), from *Microclimatology in Tropical Agriculture* (see page ix). The dashed line in (b) shows the response for groundnut, irradiance not limiting, from Ong *et al.*, 1985.

temperature is between 9–13°C, and the optimum is between 30–35°C, as for developmental rate (Chapter 1). However, an unusually high base, around 20°C, was estimated for cowpea in Nigeria[1] – much higher than the base for germination in the same species.

Extension in leaves is determinate: the duration of extension responds as a developmental period (equation 1.2), and size is determined by equation 2.4[1]. Generally, leaf size is little affected by temperature, as in the controlled environment work with pearl millet referred to in this chapter, and for maize leaves[4] and tea shoots (Table 2.1), both grown at several altitudes in western Kenya. The base temperatures for 1/duration were not estimated in these examples, but the implication is that they were similar to the corresponding ones for rate. Weak responses of leaf size to temperature have been found however. Cowpea leaves, at Ibadan in Nigeria[1], that expanded during warmer periods of the season, were slightly larger than those that expanded during cooler periods. This response is consistent with the extrapolated base temperature being about 3°C greater for rate than for duration. A similar effect of temperature on leaf size was found over a range of altitude for cassava in Columbia[5], though base temperatures were not measured. Nevertheless, leaf size was not strongly affected by temperature in any of these studies.

Table 2.1. Altitude and attributes of shoot extension for tea in western Kenya. Attributes measured when shoots reached pluckable size of two leaves and a bud. Length was significantly shorter at 2180 m than at the other altitudes (for which length was not significantly different), probably because a dry period reduced the thermal extension rate.

altitude (m)	duration (d)	rate (mm d^{-1})	length (mm)
1860	63	1.87	118
1940	69	1.57	108
2120	77	1.44	111
2180	79	1.22	96

Source: Reference 6.

Differences in the size of leaves are more likely to be caused by corresponding differences in the thermal rate or duration than by the base temperatures. The maximum size of newly expanded leaves increases with ontogenetic position for several weeks after emergence (e.g. for many cereals and legumes), or several years (e.g. for palms). Thereafter, the size of new leaves either changes little or, more commonly, decreases. This decrease might be relatively small for cereals and grain legumes, but large for crops such as cotton[1], and very large – up to sixfold – for cassava[5].

These ontogenetic changes of leaf size are sometimes caused by corresponding changes in the duration of expansion, as in cotton[1], but mostly by changes in the thermal rate. In the study of the cowpea in Nigeria[1], for example, the thermal rate for area increased from $0.98 \text{ cm}^2 \, (°\text{Cd})^{-1}$ at node 2, to $5.5 \text{ cm}^2 \, (°\text{Cd}^{-1})$ at node 8, then decreased to $3.4 \text{ cm}^2 \, (°\text{Cd})^{-1}$ at node 12. Such differences in the thermal rate are probably related to changes in the structural properties of the cells and cell walls, but might be influenced by the supply of nutrients and assimilate, which is considered later. The mean thermal rate for leaves also varies greatly between the different genotypes of a species – up to a factor of two in pearl millet, for example[3] – though thermal rates are usually greater for cereals than herbaceous legumes and other C3 plants.

The number of expanding units

Most leaf primordia are produced after seedlings have emerged. The number of leaves extending at any time on a main culm or branch is approximately the value of the thermal duration for expansion of a leaf $\hat{\Theta}_{1e}$, divided by that for initiation, $\hat{\Theta}_{1i}$. The ratio, $\hat{\Theta}_{1e}/\hat{\Theta}_{1i}$, for an apex is between 2–3 for a legume such as cowpea[1]. The number of expanding leaves for many cereals is also in this range, though initiation and rapid expansion are separated in time by the period of slow extension before the leaf 'appears'.

The number of expanding leaves increases more rapidly when branching begins on the main stem above the ground. The time from sowing to the appearance of the first tiller or branch can be defined by a thermal duration and base temperature (equation 1.2, Chapter 1). For example, in both pearl millet and groundnut, the first tillers were produced 200°Cd after sowing, and subsequent tillers in millet after a further 80°Cd, and additional branches in groundnut after 100°Cd. At first, leaves are produced on these branches at a similar rate to that on the main stem, so once branching begins the number of laminae increases almost exponentially with thermal time (Fig. 2.5). In some species, however, branching is more complex than this. In cassava, for example[5], the optimum temperature for branch production – 24°C in some cultivars – is much lower than for leaf production. Temperature also has diverse effects on the number of branches per branching point, depending on the genotype. One genotype produced the same number per branch point over a range of temperatures; another produced more branches per point as temperature rose; whilst yet another produced fewer.

The number of leaves on a plant increases in this exponential manner until the leaves of adjacent plants begin to mingle. Some branches then cease extending, or at least extend very slowly, with the result that the

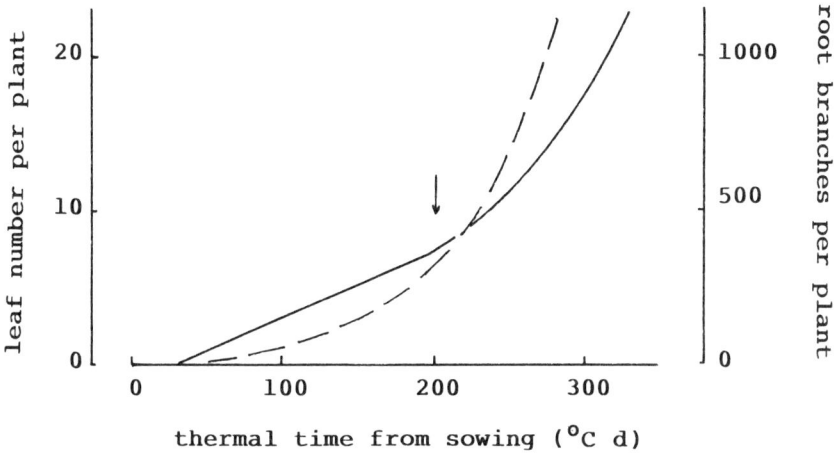

Fig. 2.5. Numbers of leaves (———) and root branches (– – –). For pearl millet
(hybrid, BK 560), grown in soil columns. The arrow shows when the first tiller was produced.
Sources: Gregory, 1983, Ong, 1983b.

number of growing points is conserved, while the number of leaves produced on a plant increases more or less linearly with time. At this time, the
number of growing points on most herbs varies inversely with the population
density over a wide range[7], but also depends on the growth habit of the
genotype. At a low population density, for example, profusely tillering
cereals and spreading legumes will have more apices per plant than more
compact forms.

Canopy expansion rate

Rapid expansion of the canopy (phase II) begins when branching raises the
number of active apices, and ontogenetic change brings about more rapid
expansion of new leaves. The little evidence available suggests that, during
phase II, the expansion rate of canopies responds to temperature much as in
equation 2.3. For example, pearl millet and cowpea canopies had base temperatures of about 10 and 20°C, respectively, and thermal rates, for leaf area
index, of 0.02 and 0.045 $(°Cd)^{-1}$. Between temperatures of 25–30°C,
canopies of the two species expanded with similar rates, despite the very
different values of the governing attributes. The millet (and presumably the
cowpea) responded in this way because temperature had little effect on the
number of apices or laminae that were extending rapidly during phase II.
Though more millet tillers were produced at lower temperatures, only two
per plant extended fully over a wide range of temperature, giving a stand of

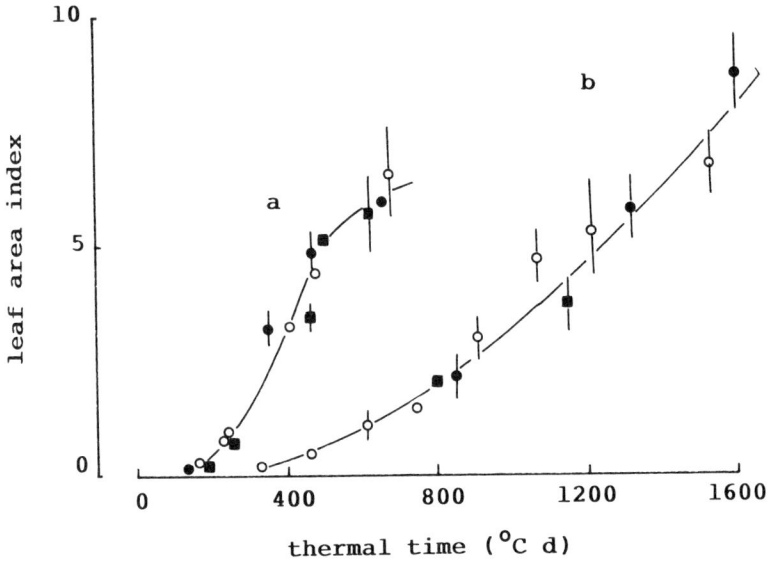

Fig. 2.6. Canopy expansion in thermal time. Leaf area index ±s.e. of (a) a short-season pearl millet and (b) groundnut, at 22°C (■), 25°C (○) and 28°C (●). Sources: Squire, 1989a and Chapter 3, Reference 8.

about 90 culms m^{-2} (each with two to three leaves expanding simultaneously during much of phase II).

Provided all the processes, including emergence, leaf initiation and leaf expansion rate, are governed by a similar base temperature, the rise of leaf area index can be expressed in relation to the passage of thermal time from sowing. This is a useful analysis for comparing canopies whose base temperatures are similar (the millet and groundnut, for example, in Figure 2.6). These canopies were established at the same population but the groundnut expanded more slowly in thermal time because its thermal durations were greater and thermal rates smaller. Dense plantings of most crops can produce canopies at rates within the range shown in Figure 2.6. Most stands however, are established at densities that are known to give maximum reproductive yields – densities usually lower than those required for the most rapid expansion. (This will be discussed in later chapters.) So, for example, the differences in expansion rate between typical stands of sorghum, cassava and oil palm (which require several weeks, several months and two or three years, respectively, to achieve good ground cover), are mainly the result of corresponding differences in the established population (e.g. 10^5, 10^4 and 10^2 ha^{-1}, respectively).

Duration and size in determinate canopies

The rate of canopy extension generally differs little between early and late genotypes of a species, and between stands in different photoperiods. Most differences in canopy size among stands of a certain type (e.g. cereals, legumes, or root crops) are caused by the factors controlling the period in which there is net gain of leaf. Among determinate plants, this period usually ends just before flowering on the main culms and branches[8]. In single-culmed cereals, expansion ceases because all the leaf primordia have expanded. In many-tillered cereals, expansion ceases even though many primordia may be unexpanded or partly expanded. In the more or less determinate legume canopies, it ceases usually because loss of old leaf overtakes gain of new leaf. For all but the single-culmed cereals, it seems that the end of expansion is controlled hormonally by other developmental events, especially those associated with reproductive growth.

The duration of expansion is therefore governed by temperature and photoperiod much as the time from sowing to flowering (Chapter 1). The delay in time-to-flowering in late, compared with early, cultivars can cause as much as a doubling of leaf area[9]. The delay caused by long days (for short-day plants) can also have major effects. The sorghum in Nigeria, referred to in Chapter 1, achieved a leaf area index greater than four when grown in long natural photoperiods, and about one in short natural photoperiods, though other factors might have contributed to this difference[7]. Generally, up to a twofold range of L is obtained by subjecting stands at one field site to a range of artificial photoperiods[9].

Effects of temperature on canopy size have rarely been thoroughly studied in the field. If the period of expansion is sensitive only, or mainly, to temperature (not photoperiod), canopy size is probably governed by a simple determinate response, similar to that for single leaves. If the base temperatures for rate and duration are similar, the effects compensate so that size changes little with temperature. The response to temperature might not be so simple if the period of expansion includes a phase whose length is insensitive to temperature or is modified by photoperiod, as in many cereals and legumes (Chapter 1). However, altitude had little effect on the leaf area index of the maize in western Kenya, implying that temperature governed expansion mainly by the determinate response in equation 2.4[10].

Duration and size: indeterminate canopies

Plants with indeterminate habit generally continue producing new leaf until the meristematic tissue is destroyed by pests, diseases or drought, or until the stand is harvested or felled. For annuals and biennials, leaf area index commonly reaches a maximum well before harvest, as loss equals or overtakes gain. This occurs even for healthy stands, usually shortly after flowering, through an effect on the senescence, or size, of leaves. This will be considered later. In perennial plantation crops, an equilibrium leaf area index can be maintained for many years, sometimes by pruning old leaves.

Temperature governs leaf area index (L) of many indeterminate canopies by a response similar to that in equation 2.5, in which L is directly proportional to $T-T_b$, at least up to T_o. In simple canopies, such as those of groundnut[11] and other indeterminate legumes, T_b and T_o are similar to those of leaf initiation, leaf expansion and branching, and are typically 10°C and 30°C, respectively. The response is more complicated in the cassava referred to earlier. Neither the base temperature nor the shape of the response was defined, but the base was probably similar to that of developmental processes. The optimum, 24°C, was much lower than that for leaf development, and was the result of different responses to temperature by the rates of leaf production, leaf expansion and branching, by the number of branches from each branching point and by leaf longevity[5].

The effects of photoperiod on the duration of expansion in indeterminate plants are less well documented than in cereals. For a long-day plant such as cassava, extending the day-length by artificial lighting in the field reduced the times to flowering and branching and increased the number of apices, but also extended the period of net leaf area gain[5]. The result was similar to that for the cereals, in that photoperiods farther from the critical increased the duration of expansion, and thereby increased L.

Nodes and internodes

Many of the effects of genotype, temperature and photoperiod on canopy size influence the number of leaves initiated on the main stem and branches (n_s in equation 2.1). These changes in leaf number are themselves associated with changes in the number of internodes and the length of stems, and in the potential number of reproductive sites in those species that bear flowers in the leaf axils.

Two main forms of response govern leaf number. In one form, the period during which the plant initiates leaves (t_s) is governed by temperature as in equation 1.2. The effect of temperature on leaf number is analogous to the determinate response in equation 2.4: n_s is given by the thermal duration for the period of leaf initiation divided by the thermal duration for initiation of a single leaf. Provided the two base temperatures are similar, n_s is independent of temperature. Leaf number is probably governed in this way in some determinate legumes, and in cereals that have a weak response to photoperiod.

In the second form of response, the period (t_s) in which leaves are initiated remains generally unaffected by temperature. The time for initiation of one leaf is still governed by equation 1.2 with a base temperature and thermal duration, $\hat{\Theta}_1$, so leaf number increases with temperature (T) above a base as in

$$n_s = t_s(T-T_b)/\hat{\Theta}_{1i} \qquad (2.6)$$

This relation governs leaf number in many genotypes. In indeterminate types, t_s is simply the time from sowing to harvest. In most cereals, t_s is governed by photoperiod, and is larger when photoperiod is nearer the ceiling value than the critical, and in late rather than in early genotypes[9]. However, there is little information on the control of leaf number in genotypes for which t_s is governed by both temperature and photoperiod. (The effect of temperature on n_s in equation 2.6 sometimes has little effect on canopy size. In the example of maize in Kenya, cited earlier, the few extra leaves produced at the highest temperature were all very small – a result of their ontogenetic position.)

These two forms of response also influence the height of the canopy (or at least the length of the stems within it). When periods for initiation of a single leaf and internode, and a whole series of leaves and internodes, are both governed by temperature above the same base, temperature should have little effect on the final number of internodes and, therefore, the length of the stem. When only the period for initiation of a single leaf is affected by temperature (i.e when the number of leaves is governed by equation 2.6), the number of internodes and the length of the stem increase with rise in temperature between T_b and T_o. When the second form of response governs leaf number, stems are longer in long days than short days (for short-day plants). When either form governs, stems are longer in late than early cultivars of a species.

Longevity, senescence and decline

This section examines the duration of the periods III and IV in Figure 2.1, and the factors responsible for restricting the rise of leaf area index in many stands and causing its eventual decline.

Duration of a canopy

Temperature has an important effect on the length of time for which stands in moist conditions maintain a canopy with sufficient leaf area to cover most of the ground. The response is similar to either the determinate or indeterminate condition described in Chapter 1 (p. 11). For the determinate, the period between the end of expansion and harvest is similar to that between flowering and grain maturity; it is therefore inversely proportional to $T-T_b$, as is the period of expansion itself. Ground cover is therefore present for the shortest time at the optimum temperature (usually 30–35°C), and for the longest time at the lowest temperature that allows germination of seed in sufficient quantity to establish a dense stand (usually several degrees above T_b). The response to temperature will be modified slightly for those canopies whose early development is also sensitive to photoperiod.

For indeterminate plants, the duration of the stand is limited by factors other than temperature, so the duration of full ground cover increases with rise in temperature between T_b and T_o. The implications of these contrasting responses for dry matter production by determinate and indeterminate stands are examined in Chapter 3.

Senescence and decline

The loss of leaf (phase IV in Fig. 2.1) can result from either, or both, of two processes in healthy stands. The most common process is senescence. At least some leaves on most annuals, and all leaves on perennials, live for a much shorter period than the canopy as a whole. The systematic effects of temperature on senescence are still uncertain. Generally, leaves remain green longer at a lower temperature, and sometimes the period of greenness responds as a developmental duration, as in cowpea[1]. This effect of temperature can be very large: for example, leaves of cassava in the altitude trial in Columbia remained on the plant for two months at 28°C and five months at 20°C[5].

The influence of temperature on senescence is probably indirect, in that the senescence coincides with another developmental event, itself controlled by temperature. Sometimes senescence begins shortly after flowering and increases with the movement of nutrients from the leaf to the fruiting structures[1,12]. Senescence also occurs in some species when the light falling on the leaf is reduced to a critical low level by shading from newer leaves, as in cassava[5] and rice[13].

The second process that reduces L is related to ontogenetic change in the size of individual units. For stands such as cassava[5] in which leaf size decreases with ontogeny, the mean area of the leaves in the canopy inevitably decreases with time. When the rate at which new leaves are produced is similar to the rate at which old leaves fall from the plant, leaf area index must decline as the stand ages.

In most investigations, the different effects of temperature, light and nutrients on L during phases III and IV have not been defined quantitatively. However, two examples illustrate the consistent effects of certain variables. The first example is for cowpea in Nigeria[1], for which temperature influenced the times to the end of leaf expansion and beginning of leaf senescence, as well as the rates of expansion and senescence. The different effects of temperature generally compensated, such that the maximum L of most of the crops was not strongly affected by temperature – a result similar, despite the senescence, to the simple determinate response for single leaves

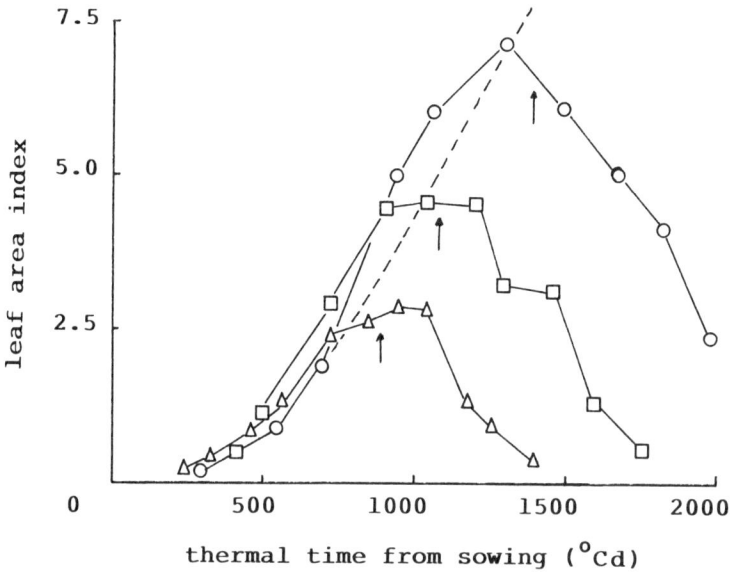

Fig. 2.7. Effects of altitude on leaf area index of groundnut. In Zimbabwe, at temperatures 23.2°C (○), 20.1°C (□), 17.9°C (△); thermal times based on air temperature above T_b = 10°C. The arrows indicate 110 d after sowing. Cultivar, Makula red. Source: Williams *et al.*, 1975. The dashed line represents the relation for groundnut in Figure 2.6.

and determinate canopies. The second example is for rice[13]: among several cultivars, the maximum L was larger for canopies whose architecture allowed more light to be transmitted to the lower leaves, i.e. maximum L was inversely related to the extinction coefficient (defined in Chapter 3).

The control of L is usually less clear than in these examples. As a case of greater complexity, Figure 2.7 shows L for three crops of groundnut grown at different altitudes in Zimbabwe. Leaf area index increased more rapidly in warmer conditions: the response appeared similar to that defined by equation 2.3. Thermal time was calculated assuming a base temperature of 10°C, which was measured for groundnut in controlled environments. The relation between L and thermal time was initially similar at the three sites, and similar to that for the groundnut in Figure 2.6. Later, loss of leaf became greater than gain. The maximum leaf area occurred at a similar calendar time at the three sites. The factor causing the onset of senescence is not known and might not have been related to development. The rate of decrease in L also responded to temperature in much the same way as the rate of increase (i.e. it was faster at higher temperature). Therefore, L declined at about the same rate in relation to thermal time at the three altitudes. Despite the senescence, both the maximum L and the duration of a sizable canopy still increased with rise in temperature above T_b – effects typical of the general indeterminate condition (e.g. equation 2.5).

Temperature and determinacy: summary

The strong responses to temperature of the rates and durations of many processes of development, expansion and senescence have an important bearing on the size and longevity of canopies. For determinate canopies, temperature has little effect on maximum leaf area index, but has a strong effect on the duration of leaf area – which is greatest at the lowest temperature enabling survival of most seeds (typically 15–20°C). For indeterminate canopies, temperature affects both maximum size and duration: both attributes are largest at the temperature at which the canopies expand most rapidly (typically about 30°C).

The leaf canopy: effects of other (mainly restrictive) factors

Many of the responses described in the previous section are influenced by other environmental factors, principally drought, nutrients and solar radiation. These factors generally affect rates more than durations.

Solar radiation

It was concluded in Chapter 1 that variations in the supply of assimilate, as influenced by the natural variation in solar radiation, have little effect on the length of developmental periods. In moist conditions, assimilate also seems to be produced by photosynthesis generally faster than required by expanding laminae. In Figure 2.4b, for example, the rate of expansion was restricted only when solar radiation fell unnaturally low (about 10% of full tropical sunlight) during the middle of the day. This is consistent with the fact that the leaf area per plant of a shorter partner in an intercrop is commonly little affected by the low irradiance it receives, which is typically 30–50% of full sunlight if the taller partner is a cereal. Low irradiance in canopies is more likely to affect the longevity of leaves, as described earlier.

Other studies have shown that shading canopies can have a small or moderate effect on leaf area. In an experiment in India, stands of groundnut in the field were covered with artificial 50% shade to simulate the effect of a taller cereal in an intercrop. The shade slightly reduced leaf size, the rate of branching and, thereby, total leaf number. It also brought forward senescence by a few days and so reduced the duration of expansion. These effects together only reduced L to 90% of the value in the unshaded control[14]. In Queensland, Australia, L for cassava was hardly affected by shade that reduced irradiance to 80% of full sunlight, and L was only reduced to 73% of the unshaded when irradiance was reduced to 32%[5].

Less is known of the effect of low irradiance on the size of stems in the field. During expansion, stems require dry matter at a much faster rate than leaves. The thermal duration of expansion would probably be unaffected by assimilate, so stems would still expand for a shorter period as temperature rose; but the thermal rate could be restricted, especially at temperatures near the optimum. A limit imposed on the rate would then cause stems to be smaller at a high, as opposed to a moderate or low temperature. This effect can be illustrated by redrawing Figure 2.3a with the two durations remaining unaltered (i.e. duration longer at the cooler than the warmer temperature), but with the rates the same at both temperatures. The effect might have contributed to the variation with altitude of stem height of the maize in western Kenya described earlier[4]. Stems were shortest at the warmest temperature, despite the few additional internodes at this temperature caused by the response in equation 2.6 (see also Fig. 1.7).

Drought

Many stands in the tropics experience a period, even if only a few days, in which dry soil or dry air limit the processes controlling the size of canopies.

Emergence and establishment

In the drier parts of the tropics, a severe environment near the soil surface can prevent most seeds from germinating and developing into strong seedlings. If this happens, there might be too few individuals to support a canopy and root system that will efficiently utilize the available resources. A dry soil, particularly one with a surface 'cap', might restrict penetration of the plumule and radicle, either by reducing the turgor potential of the meristematic tissue, by increasing the force these structures have to exert to push through the soil or by restricting the supply of oxygen to the metabolizing tissue. As for emergence in moist conditions, there has been little systematic analysis in the field to distinguish the relative importance of these factors.

These factors probably do not alter the base temperature for development, but they may increase the thermal duration. Accordingly, thermal time analysis is ideal for defining effects of soil dryness in the natural environment where the temperature is changing[15]. A full analysis should take account of the effect on the spread of thermal time for emergence within the population and on the fraction of the sown population that emerge, as shown for depth of sowing in Figure 2.2.

Responses of leaf size and number

The rate at which existing leaves expand is usually the first process controlling canopy size to be affected by dry conditions. The duration of expansion
is more stable, so the reduction of rate causes a corresponding reduction in
leaf size. Dry air and dry soil seem to have independent effects.

Dry conditions could reduce expansion rate through several physiological mechanisms. The turgor potential of the leaf cells is commonly reduced, and restricts extension rate by a linear response (Fig. 2.4b), which
can be written as

$$\delta l/\delta t = \sigma(\Psi_p - \Psi_{pb}) \tag{2.7}$$

where σ is the rate of extension per unit of turgor potential above a base turgor potential, Ψ_{pb}[3,16]. The corresponding relation with leaf water potential
is usually not so tight. A fall in the water potential, during the day or through
a drought, is often compensated by a fall in the solute potential which conserves the turgor potential[16].

Species differ much in the extent to which dry air and dry soil reduce leaf
turgor and thereby extension rate. Of the two species in Figure 2.4b, the millet loses turgor during the day much less than groundnut. The millet extends
most of the time, whereas the groundnut sometimes extends more during
the night than the day, despite the lower temperature at night. In such
circumstances, it is sometimes unclear whether temperature or turgor
primarily limits extension rate, and whether the response to one of these
factors is influenced by the value of the other[16].

Nevertheless, a response such as that shown by equation 2.7 offers
simple attributes for characterizing a species. In Figure 2.4b, the millet
leaves extended twice as fast as the groundnut because of a difference in σ;
the base turgor potentials were similar for both species. Generally, these
attributes seem less conservative than those governing the response to temperature. Sometimes dry conditions reduce extension rate, yet have no
effect on the water potential of the whole leaf or a substantial part of it[5].
Admittedly, in some of these instances, drought might still be affecting the
turgor potential in the small area of meristematic tissue, but there are probably
mechanisms that control extension rate independently of turgor. Some
effects of dry soil might restrict the availability of nutrients to the plant. Any
reduction in the rate of photosynthesis might restrict extension by providing
insufficient assimilate to the new tissue, especially if the plants are small and
have little dry matter in store. There might also be hormonal responses to
dry soil that influence the extensibility of cell walls independently of the
turgor of the leaf (see Further reading, this chapter).

During a mild drought, leaf size might be the only gross morphological
attribute affected. If a drought becomes severe, the rates of leaf initiation

and branching are both eventually reduced. Leaf life is also commonly shortened (though not always in cassava, for which the reduction in the size of
newly expanded leaves allows more light to fall on the older lower leaves).
Ultimately, whole branches or tillers cease developing, or die if already initiated. This restriction of branching conserves the rate of initiation on the
main branch to an extent. Dryland cereals, for example, maintain very conservative initiation rates and leaf numbers on their main culms (Chapter 1).
If, however, the stand uses most of the water available to it, development
ceases in many individuals. Leaf desiccation follows and some plants die.
These responses have seldom been quantified in terms of any physiological
index of dryness.

Despite these several effects of drought on the size and number of leaves,
expansion of leaf area at the scale of the whole plant or stand is still sometimes related to the leaf water potential. Figure 2.8 gives an example, in
which expansion is expressd as a thermal rate, to account for small differences between treatments in leaf temperature. The difference in L between
a well-watered and a water-stressed stand is then related to the mean, or accumulated, difference between them in leaf water potential (perhaps measured once or several times a day). This was found for cowpea in a wide range
of treatments in California, USA[17]. Here at least, a response of the type
shown in Figure 2.8 seems to have been conservative over several months.
But this conservatism is probably rare.

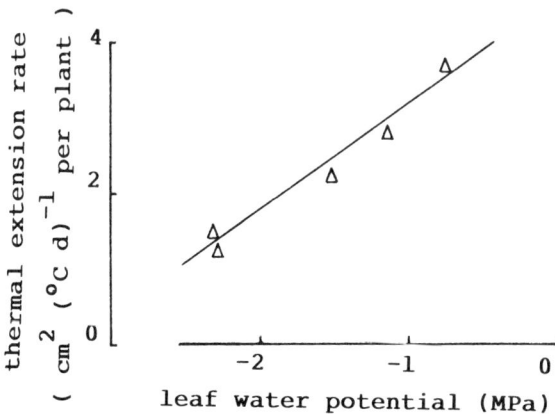

**Fig. 2.8. Relation between leaf area expansion and water
potential of groundnut.** For stands of cv. Kadiri 3
with their roots in field soil in controlled environment
glasshouses, water potential measured on expanding leaf
between 0800 and 0900 GMT, adapted from Stirling, Ong
and Black, 1989.

Survival, and expansion after drought

Desiccation and death is the fate of many dryland crops if the rains fail, but a drought during the growing season is a feature of those climates that have a bimodal rainfall distribution. Stands in these conditions have the opportunity to continue expanding when the drought is relieved. Whether they do continue depends much on the state of the foliage and meristems at the end of the drought. Most dryland crops, including sorghum, cowpea and groundnut, can keep meristems alive during a long drought, even if much of the leaf dies[18].

Not all cultivars of such species have this ability to survive, but those that have share several common attributes. In a comparison of sorghum cultivars at Hyderabad in India[18], resistant cultivars survived a severe drought with much of the leaf still green. When there was further rain, new leaf matter expanded to produce a large canopy and the stands yielded grain. Other more susceptible cultivars survived, though with little leaf; they failed to take advantage of the rain and yielded nothing. Compared with the susceptible strains, the more resistant had the following attributes: slower expansion of both canopy and root system; relatively faster expansion of the root system than the canopy, giving rise to a larger root/leaf ratio; faster transpiration per unit leaf area (Chapter 4); a more hydrated leaf tissue (higher water potential) and a greater degree of leaf rolling to avoid direct solar radiation. The difference in leaf rolling was related to the water potential to which leaves had to be reduced before rolling occurred. In the most resistant, rolling occurred when the potential was about -2 MPa; in the least resistant, it hardly occurred, though the potential fell below -3 MPa (Fig. 2.9).

In a comparable study with groundnut at the same site[18], resistance to drought was associated with a different set of attributes. The most successful cultivar produced the most leaves (unlike the comparable sorghum), and, thereby, the most leaf axils from which the reproductive sinks originate. It also held its leaves the most vertically and was best able to fold its leaflets during drought, attributes that enabled it to produce more potential reproductive sinks without having to intercept more solar radiation than the other cultivars.

A further characteristic of successful dryland plants is that their canopies suffer virtually no after-effect of the drought. A few days after rain, they expand at a rate, and photosynthesize with an efficiency, little different from that of new canopies.

The relevance of dry air

In dryland agriculture, both dry air and dry soil usually act to reduce leaf area. Their separate effects are difficult to distinguish. In irrigated agriculture, dry air alone can greatly reduce expansion. The degree of dryness

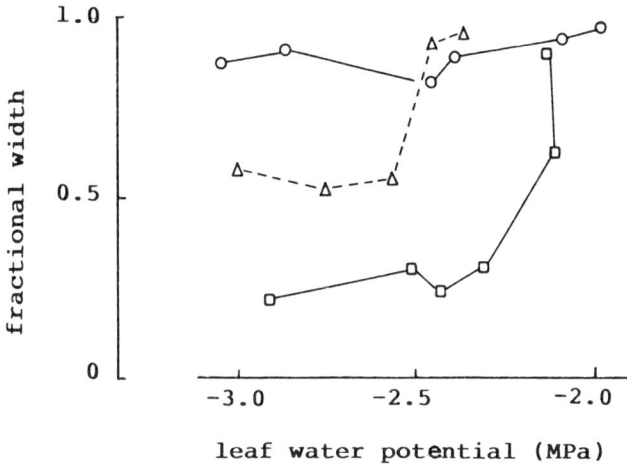

Fig 2.9. Genotypic differences in the water potential required to cause leaf rolling in sorghum. The vertical axis represents the width of leaves as a fraction of their unrolled width. The lines show the relation for genotypes differing in the degree of resistance to scorching: □, most resistant; △, intermediate; ○, least resistant.
Source: Matthews *et al.*, 1989b.

of the air is best expressed as the 'saturation water vapour pressure deficit', shortened to saturation deficit, $D^{[19]}$.

Field work on tea in Malawi revealed one of the first quantitative relations between expansion rate and $D^{[6]}$. Extension of the shoots was measured with a ruler and related to temperature and D recorded in a meterological screen. The length of the shoots increases approximately exponentially in time, and can be expressed as a relative extension rate (with units such as cm cm^{-1} d^{-1}). This rate is governed by temperature as is the linear extension of laminae (in Fig. 2.4a), and the response can be defined by a base temperature and a thermal (relative) extension rate as in equation 2.3. Since temperature and D both changed with the seasons, the effect of D could only be examined using the thermal rate. This form of analysis is comparable to that in Figure 2.8, and has the same purpose as thermal time analysis of developmental periods (Chapter 1), i.e. the response of a process to another limiting factor can be examined even when temperature is changing.

The relation shown in Figure 2.10 suggests that the rate was little affected by change in D below 2.3 kPa, but decreased linearly with rise in D above 2.3 kPa. These results explain why the yield of tea responds well to irrigation in a cool dry season, when the saturation deficit is small, and sometimes responds weakly in a hot dry season, when it is large$^{[6]}$.

The effect of D on the thermal rate has since been investigated for other

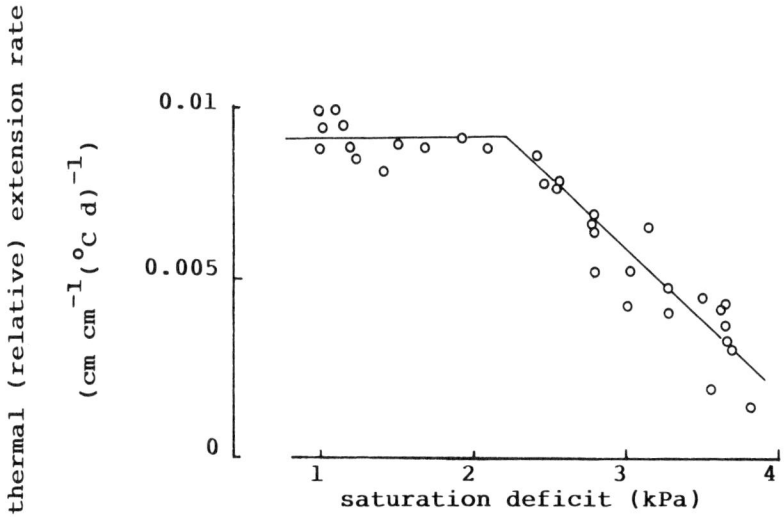

Fig. 2.10. Relation between shoot extension rate (tea) and saturation deficit in Malawi. The thermal rate is from weekly measurements of shoot growth and mean temperature ($T_b = 12.8°C$). The saturation deficit is the corresponding mean at 1400 h. See text for further information. Adapted from Tanton, 1982b.

species in realistic controlled environments[3]. The form of the response is inconsistent however: even for the same cultivar, the thermal rate sometimes decreases as D increases, but at other times it changes little over a wide range of D. The reason for this inconsistency has not been found, but D could have several effects on extension, some or all of which might not operate in certain circumstances. As D rises, the transpiration rate also commonly rises, and drives down the leaf water potential. Saturation deficit probably acted in this way on the tea: the leaf water potential decreased when D was large, even though the soil was thoroughly moist. The low water potential probably had an effect such as that in Figure 2.8, restricting extension by reducing the shoot turgor potential (cf. Fig. 2.4a). In other circumstances, the water potential might change little if the leaf conductance decreases in response to D, and so conserves the transpiration rate (see Chapter 4). This decrease in conductance might also cause a decrease in the rate at which new assimilate is produced by photosynthesis, and this effect might itself restrict extension.

Nutrient shortage

The systematic responses of leaf area to nutrients are not well understood. Adding nutrients to the soil usually increases the size of the canopy and affects senescence, though inconsistently. The size is influenced mainly through the rate of extension; the duration is little affected, or is sometimes greater, when nutrients are scarce than when they are plentiful (Chapter 1).

Senescence is associated with, and probably caused by, the movement of nutrients from leaves to fruiting structures[12,20]. The amounts of nutrients moved, and the degree of leaf senescence, probably depend on the extent to which demand for nutrients by the fruits can be satisfied by the supply of nutrients from sources other than the leaves. Senescence is therefore influenced by the size of the reproductive sink for nutrients, by the quantity of nutrients taken up from the soil during fruiting and by the quantities stored in other vegetative structures before fruiting begins (later moved from these structures to fruits). Adding fertilizer can affect all these factors and might delay or hasten senescence, depending on the timing and magnitude of the different effects.

Nitrogen content and leaf expansion rate

It has been difficult to develop relations between leaf nitrogen content and the expansion rate, or size, of leaves and canopies. One reason for this is that nitrogen can affect expansion through several physiological processes, including cell division at the meristems and photosynthesis in the expanded laminae. Some of these processes might be influenced less by the concentration of N in the bulk of the leaf than by the local concentration, which is not usually measured. Another reason is that the nitrogen concentration might have little relevance if other factors, including the concentrations of other nutrients, limit expansion. A third reason is that, in some species, new tissue expands at a rate which conserves the nitrogen content per unit leaf area. When this happens, the leaf area of the canopy increases if more fertilizer is given, while the nitrogen content per unit leaf area changes little[5].

Even when treatments affect both leaf nitrogen content and leaf expansion, the relations between them are not straightforward, as they are greatly influenced by the plant population density. As an example, Figure 2.11 shows how expansion rate was related to N-content per unit leaf area in stands of maize in Zimbabwe, and stands of maize and sorghum in Australia[20]. In Zimbabwe, the experimental treatments were five different population densities (2.3, 3.5, 4.8, 6.1 and 7.4 m^{-2}) and fertilizer was applied in all at a rate judged not to limit growth at the highest population. In Australia, the treatments were five different levels of applied nitrogen, but the population was higher for sorghum ($16\,m^{-2}$) than for maize ($7\,m^{-2}$).

The N-content per unit leaf area increased as the amount of fertilizer was increased (since more fertilizer was shared between the same number of plants) and also increased as the plant population was reduced (since a similar amount of fertilizer was shared between fewer individuals). The relation between the canopy expansion rate, $\delta L/\delta t$, and N-content was different for the three stands (Fig. 2.11a). In Zimbabwe, δL/δt decreased with a rise in N-content, because of the concomitant rise in population. In Australia, δL/δt increased (probably asymptotically) with increase in N-content, but was greater for the (16 m^{-2}) sorghum than the (7 m^{-2}) maize over the whole

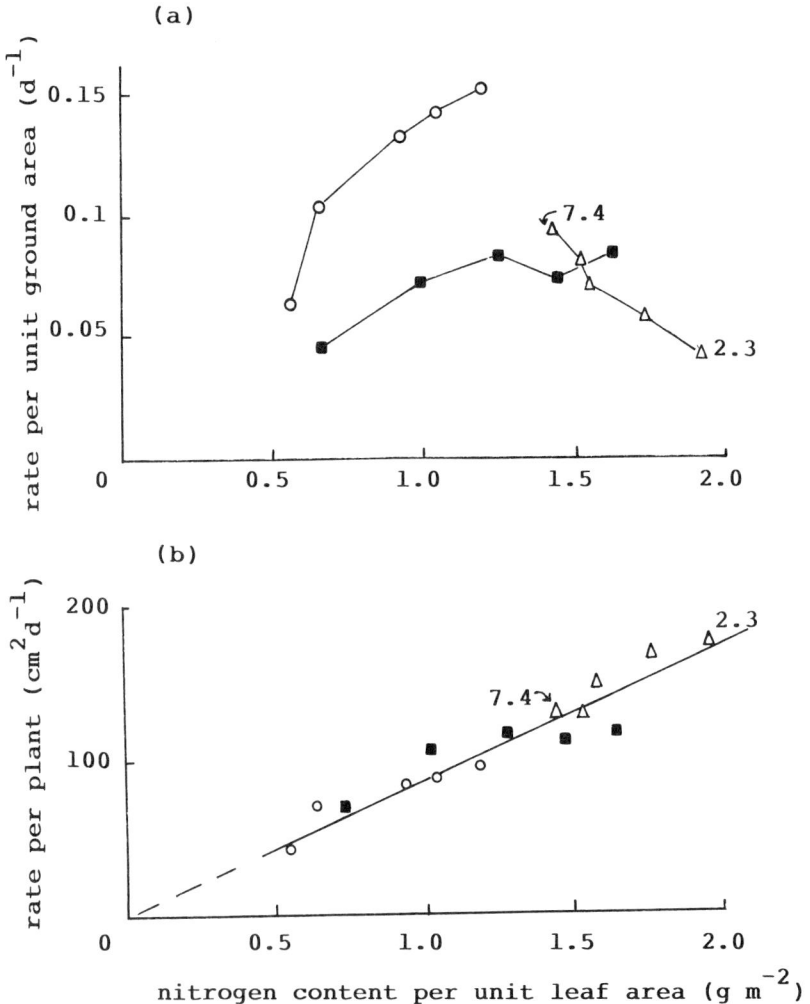

Fig 2.11. Leaf expansion rate and nitrogen content. For (a) leaf area index, and (b) leaf area per plant. Site/species/plant population/N-treatment: △, Zimbabwe, maize, 2.3–7.4 m^{-2}, uniform N application; ■, Australia, maize, 7 m^{-2}, five levels of applied N; ○, same site in Aus., sorghum, 16 m^{-1}; same N levels. Expansion rate is the average during phases I and II, i.e. 0–70 d after sowing in Zimbabwe, 0–47 d in Australia; N-content per unit leaf area is the value at the end of phase II. Numbers show lowest and highest population density in Zimbabwe experiment. The line in (b) is a linear regression for all co-ordinates: y = 1+86x, r^2 = 0.89. See text for interpretation. Sources and further information, Reference 20.

range of N-content. Despite this complexity at the level of the canopy, the underlying relation between the leaf expansion rate of *individual* plants and N-content is consistent for the three stands (Fig. 2.11b).

In these examples, it is uncertain that N-content governed the leaf expansion rate of individuals in all stands. In some, the rate might have been determined by another factor and N-content might only have been indirectly related to this factor. If N-content was the causal factor in this, or similar, relations, it could only affect the expansion rate up to the maximum rate that the genotype could achieve in the prevailing conditions. This maximum would be determined by attributes such as the maximum number of leaves that can expand at any time, and their maximum thermal expansion rate, and by any environmental factors imposing a limit on the rate, e.g. low temperature, drought or heavy shade. For the maize cultivar grown in Australia, the maximum rate appeared to be about 110 cm^{-2} d^{-1} (leaf area per plant) when the N-content was about 1.0 g m^{-2}, but for the other two stands all rates seemed to be below the respective maximum.

These results and analysis concur with the common opinion that effects on plant growth of population and nitrogen are interdependent. Moreover, they suggest that the reduction of the leaf area of individuals that occurs as plant population increases might be partly the result of there being insufficient nitrogen in the soil to allow each plant to expand at its maximum rate – even on fertile soil, and with added nutrients.

The root system

There is much less scope for examining roots and root systems in a way comparable to that presented for leaves and canopies. Measurement of root length is tedious and laborious, and so there have been few detailed studies for stands in the tropics. Moreover, many laboratory experiments on rooting have confined the roots to unreasonably small volumes of soil. The following account relies heavily on several field studies in India, and a few experiments with roots grown in soil columns.

Root elements: experiments with soil columns

Temperature affects the extension rate of radicles much as that of laminae[3], but there is little information on the response for individual roots deep in soil. In a study in which plants of pearl millet were grown in soil columns, sunk in the earth within controlled environment glasshouses[21], the extension rate of root axes increased with rise in temperature. The relation was not as tight as for leaves, and rates also increased with time after initiation of the axis, reaching a maximum of 60–70 mm d^{-1}. The thermal rate for root extension increased with ontogeny by up to a factor of two.

Root elements can extend for a much longer period than laminae. In determinate plants, the root system usually ceases expanding at around the time of flowering (see later), so the period for which at least some roots extend might be controlled by temperature as for leaves. The size of elements would therefore be unaffected by temperature. In indeterminate plants, the size of the main roots might be governed by equation 2.5, so any time after sowing the length of the elements would increase with rise in temperature between T_b and T_o.

In the experiments with millet, the number of lateral roots, formed on the nodal root axes, increased exponentially in thermal time (Fig. 2.5). About 100 lateral branches had been formed on each plant at 150°Cd after sowing, and 1000 at 250°Cd. The response seemed to be governed by the temperature near the apical meristem, which was near the soil surface at this time, and not that of the deeper soil in which the roots were branching. There is little known of the number of simultaneously expanding nodal roots of a cereal, or main branches on a tap root, but the figure is potentially much larger than for leaves, because individual roots can extend for a much longer period.

Work in soil columns has also revealed an important response to soil dryness. Plants of sorghum were established in a moist column, which was not subsequently re-wetted. When the plants had dried the surface layers of soil, some of the root axes ceased extending when still close to the surface while the rest extended downwards more rapidly than in moist soil[21].

Expansion and size of root systems

It is not known whether expansion of root systems can be defined by a base temperature and thermal rate, but the root length per unit soil volume (l_v) increases at a rate that generally changes little during phase II provided the soil is reasonably uniform. This linearity suggests that, as for leaves, some check is imposed on the number of elements expanding at any particular time. When a soil is moist, roots proliferate most rapidly in the soil just below the surface. In the examples shown in Figure 2.12, rates of proliferation are about 0.7 mm cm^{-3} (soil) per day for cassava and pearl millet, and 0.4 mm cm^{-3} for groundnut. The rate decreases with distance from the soil surface, to an extent that l_v in a moist soil commonly declines more or less exponentially with depth (Fig. 2.13).

The rate of increase in total root length, $\delta R/\delta t$, also usually changes little with time during phase II, but can differ much between sites and species depending on the rate of extension deep in the soil. The millet and groundnut in Figure 2.12 produced most of their roots in the top 0.3 m of soil, and $\delta R/\delta t$

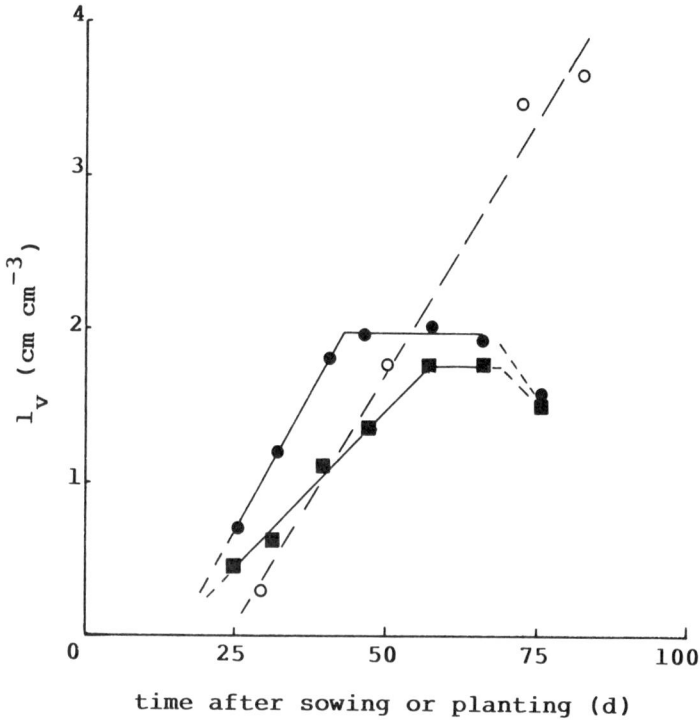

Fig. 2.12. Extension of the root system in three species. Root length per unit soil volume (l_v) in the top of 0.2 m of soil, for cassava (○), groundnut (■) and pearl millet (●). Sources: Gregory and Reddy, 1982, Aresta and Fukai, 1984.

was respectively 120 and 70 m m^{-2} (field surface) d^{-1}. The cassava produced many roots down to below 1.2 m, and $\delta R/\delta t$ was 500 m m^{-2} d^{-1}.

Differences in the duration of expansion sometimes accentuate these differences in rate to give root systems of very different size in moist soil. The root systems of determinate plants, such as the cereals, cease extending at about the time of flowering on the main culm[7]. They still have many elements capable of expanding, and the end of expansion seems to be signalled by the start of reproductive growth (cf. leaves on tillering cereals). A short-season plant, such as the pearl millet in Figures 2.12 and 2.13d (for which phase II might be 25 d in a day-length of 12 h and a temperature of 25°C) would produce a root system with a total length of perhaps 3 km m^{-2} (field). Long days might increase the total root length, though by how much is not known, and a long-season cereal could produce a much larger system than this. For example, the upland rice in Figure 2.13a produced a total root length of 24 km m^2 – expansion was more rapid and for a longer period.

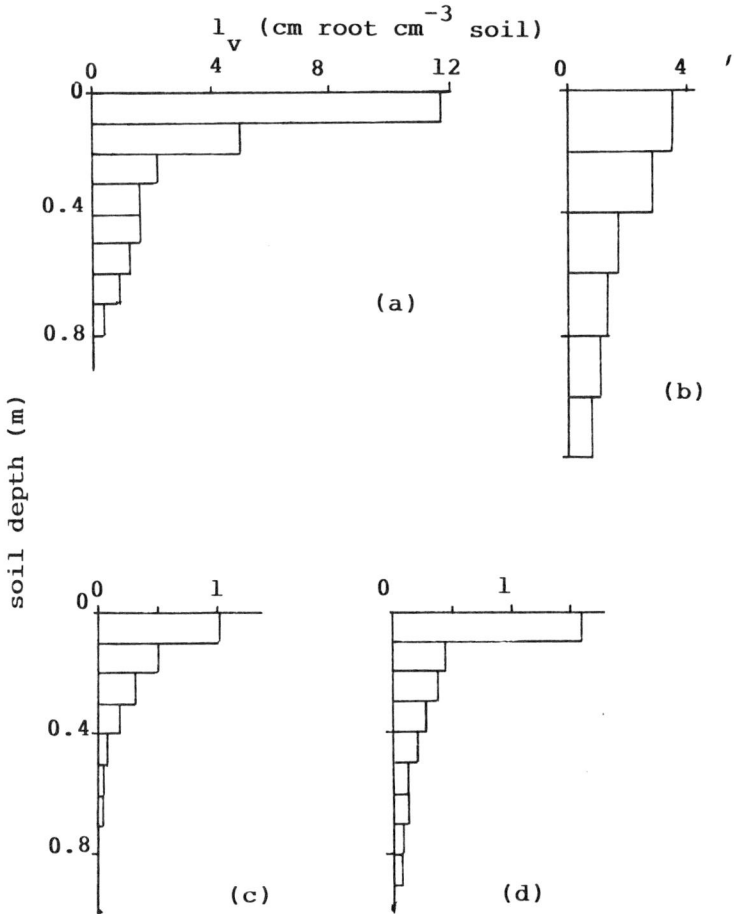

Fig. 2.13. Examples of rooting profiles in moist soil. Length of root per unit soil volume (l_v), at maximum size of the root system, for (a) upland rice, (b) cassava, (c) groundnut and (d) a short-season pearl millet. Sources: Yoshida and Hasegawa, 1982 (a), Aresta and Fukai, 1984 (b), Gregory and Reddy, 1982 (c and d). Note difference in horizontal scale between upper and lower pairs of profiles.

The groundnut in Figure 2.12 expanded for longer than the millet to achieve an only slightly smaller total length, but other legumes extending longer than this can produce root systems of up to 7 km m^{-2}. The root systems of indeterminate species, extending for several months, can equal or surpass the rice in Figure 2.13a. The cassava in Australia (Figs 2.12, 2.13b) also achieved a total root length of 24 km m^{-2} after 100 d, but differed from the rice in that root length was more evenly distributed throughout the soil profile.

The depth of the system differs among stands as a result of the differences in the rate and duration of the descent of the leading roots. Most cereals and legumes in moist soils do not produce much root below 0.3 m, though some roots will usually extend to at least 1 m. The cassava referred to produced many roots below 1.5 m, and perennials of longer duration will eventually extend their roots to depths of several metres. Genotypic differences in rooting depth have not been extensively studied, but can be large, as in rice, lowland cultivars of which descend less rapidly than upland[22].

Among most of the herbaceous stands examined in this volume, R in moist, fertilized soils differed over about a tenfold range – from around 3 km m^{-2} for the early cereals and legumes fo 31–33 km m^{-2} for paddy rice and cassava. The factors controlling the rate of extension are still largely obscure. For example, cassava stands in Columbia produced a root length thirty times smaller than those in Australia, though the reason for the difference was not discovered[5]. Effects of soil conditions, specifically those affecting the ability of roots to penetrate soil pores, undoubtedly contribute to differences in rate between stands, but have seldom been examined systematically, even in experimental systems[23].

Influences of nutrients and drought

There seems very little systematic and quantitative information on the effects of nutrients on root systems in the field. In cereals, nutrients might affect the duration of expansion through their small influence on the time of flowering (Chapter 1). Roots proliferate most where nutrients are concentrated. If fertilizer is given, l_v just below the soil surface can be much increased[24].

There is more information on the response to dry soil. The form of response depends very much on the distribution of water with depth in the soil. The response also depends on the mechanical properties of the soil, specifically the existence and extent of dense or compacted layers (which slow and might halt the descent of the root systems). The rate of expansion can be slower in a partly dry than in a moist soil. Expansion might cease if the extending root front meets a layer of dry soil – only continuing if the soil profile is later wetted at depth. Slow dry matter production during drought might also restrict the amount of new assimilate available for root growth[25].

If, however, the soil profile is fully charged with water, but receives no rain once the stand is established, roots usually descend faster than in moist soil. Proliferation of lateral roots shows a similar response to drought as that observed in the soil columns. For the millet and groundnut in Figure 2.12, the laterals just below the soil surface ceased extending after 25 d in a drying soil, and also when l_v was only half what it was in the wet soil, in which they extended until flowering at 45 d. It is uncertain why these laterals stopped extending prematurely, but they may have because assimilate was diverted to roots deeper in the soil, or because they dried the soil to an extent that prevented further extension. However, laterals deeper in the soil extended more rapidly than they would at a similar depth in moist soil, at least until they had dried the soil appreciably. Consequently, total root length increased more rapidly than it would in soil that was periodically re-wetted at the surface (Fig. 2.14).

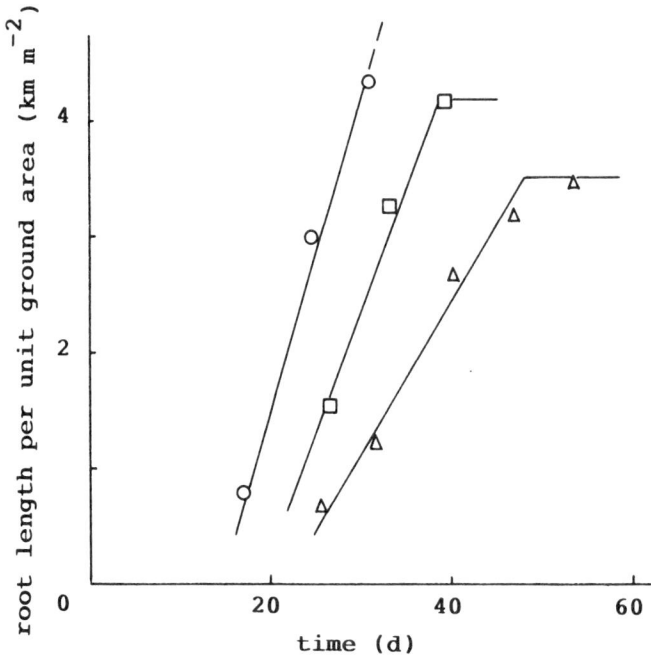

Fig. 2.14. **Total length of root per unit area of ground for stands of pearl millet.** $\circ$, Niamey, Niger, grown on stored water, daily maximum saturation deficit (D_{max}) 4 kPa (Azam-Ali *et al.*, 1984a); $\square$, Hyderabad, India, on stored water, D_{max} 2.4 kPa (Gregory and Squire, 1979); $\triangle$, Hyberabad in the monsoon season, D_{max} 1.7 kPa (Gregory and Reddy, 1982).

Relations between root length and leaf area

The work in soil columns with pearl millet, referred to earlier, illustrates how the development and expansion of leaf and root systems are related[21]. Once tillering began, the number of root elements increased more rapidly than the number of leaves (Fig. 2.5). When the first tiller appeared, about 50 lateral roots were produced for each lamina; but this ratio increased with time (at least up to 300 °Cd when the measurements of roots ceased), since roots branched more frequently than new leaves were initiated. This imbalance in number was compensated by the increase with ontogeny in the thermal extension rate and width of leaves. As a result, R/L was very conservative at about 1.4 km (root) m^{-2} (leaf) during phase I and the early part of phase II (when the measurements ceased).

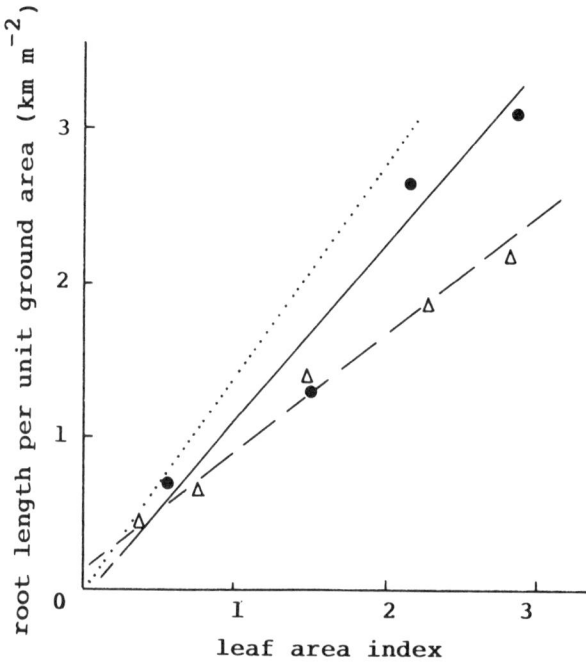

Fig. 2.15. Root length/leaf area ratios during expansion.
For pearl millet (●, ———) and groundnut ($\triangle$,– – –) during expansion in the field in India. Sources: Gregory and Reddy, 1982; Gregory and Squire, 1979. The lines are linear regressions indicating ratios of 1.14 km m^{-2} for millet (solid line) and 0.71 for groundnut (dashed line). The dotted line, 1.4 km m^{-2}, is the relation for millet in controlled environment glasshouses, from Gregory, 1986.

A similar R/L ratio was found throughout phase II for the same cultivar of millet grown in moist conditions in India (Fig. 2.15). Groundnut, grown at the same site and time, extended with a slightly smaller ratio than the millet: the slower leaf extension of the groundnut (caused by the smaller thermal extension rate) more or less matched the slower root extension. The R/L ratio can be more variable than for the millet and groundnut in Figure 2.15. Nevertheless, R/L during expansion differs between species much less than either R or L. For example, R/L during phase II for the cassava in

Figure 2.12 was 2.2 km m^{-2}, only two to three times larger than the values for millet and groundnut, despite R itself being about ten times larger.

Environmental modification of the root length/leaf area ratio

The ratio R/L changes much more than indicated above if sunlight, water or nutrients are in short supply. The plant then responds by allocating a greater fraction of the assimilate to the structure whose function it is to obtain the limiting resource[27]. For example, the shading of cassava, referred to earlier, caused the extension of fibrous roots to be reduced much more than expansion of the leaves. Root length in the 30% shade was half the value without shade and the R/L ratio therefore decreased with shade. The limiting resource was solar radiation, and the plants responded by investing more in leaf area than root length. The opposite response occurred in fertilizer experiments with barley in Syria[24]. When fertilizer was withheld, both leaf area and root length decreased, but area was reduced more than root length in relative terms, (Table 2.2) and so the R/L ratio increased. Here, the limiting resource was in the soil, and the plants invested relatively more in root than leaf.

Table 2.2. Effect of fertilizer on root length per unit leaf area. For barley at two sites (Jindiress and Breda) in Syria.

Fertilizer	R/L (km m^{-2})	
	Jindiress	Breda
None	7.7	5.1
+potassium	7.1	4.1
+potassium+nitrogen	3.5	3.9

Source: Gregory, Shepherd and Cooper (1984).

Responses to drought are qualitatively similar to those to nutrient variations. A drying soil commonly restricts leaf area and stimulates root expansion. In combination, these effects increase the R/L ratio to as much as six times the values typical of stands in moist soil. However, R/L seems more variable in dry than moist soil. It sometimes changes systematically with time. For example, R/L for pearl millet, growing on stored water in a sandy soil in Niger[26], decreased from 6 km (root) m^{-2} (leaf) at about 20 d after sowing to 3 km m^{-2} by 30 d, mainly because the rate of leaf expansion increased in the intervening period. The ratio for barley at a dry site in Syria changed similarly – from just more than 6 km m^{-2} at the beginning of rapid stem extension, to 3–4 km m^{-2} at flowering[24]. It is not clear how these large changes are regulated, but there is evidence that R/L is increased by the low population densities at which stands are usually grown in dry climates (see below).

Effects of plant population density

Effects of plant population density (N_p) on L and R have been referred to several times in this chapter. Population has little effect on the duration of phase II, so influences the maximum value of L or R at the end of phase II through effects on the rate of expansion. Tnere is no information on the relation between N_p and an attribute such as the thermal expansion rate, but it is well established that, as the population rises, the rate of expansion in phase II increases, though the effect of an incremental rise in population gradually diminishes[7].

The growth habit of the plant influences the relation between population and expansion rate. Individuals of much branched and spreading forms are able to alter the number of active apices, and can usually expand both more rapidly at low population and more slowly at high population. For these forms, expansion rate changes relatively little over a wide range of population.

Certain genotypes of rice and pigeon pea are among the most flexible in this respect, and can be contrasted with oil palm, which is one of the least flexible.

Quantitative analysis of growth habit and plasticity

The 'plasticity' of genotypes can be defined quantitatively in terms of the relation between N_p and the leaf area or root length of the individual. The most common response for leaf area is represented in Figure 2.16. At very low populations, adjacent plants have little influences on each other and individuals achieve the maximum area possible for the genotype in the prevailing conditions (a_{max} in Fig. 2.16). As N_p is increased above the value (N_{p1}) at which neighbours begin to compete for the resources, the leaf area of individuals decreases in the more or less exponential manner shown by line (i) on Figure 2.16. Very few experiments with crops examine populations high enough to cause much self-thinning, but a minimum area is shown in Figure 2.16 representing that necessary for the individual to survive (a_{min}). The response of area to N_p is similar to that of mean plant dry weight, because the area to dry weight ratio is much less sensitive to N_p than either area or weight (see Chapter 6).

For comparing genotypes, it is more convenient to use the reciprocal of mean plant area, which is related to N_p by curve (ii) in Figure 2.16. An important feature of this curve in that 1/area increases linearly with N_p over a very wide range. Most experiments on N_p remain on the linear part of the curve, and a linear regression of 1/area on N_p gives an estimate of the slope, which usually has a positive intercept on the y-axis. The slope indicates the degree to which the plant can alter its leaf area in response to change in population. A genotype with a larger value of slope than another is more able to change its area as population changes. For example, if the units of both N_p and 1/area are m^{-2}, the slope ranges from 0.17 for a very plastic pigeon pea to 0.03 for the more rigid oil palm[28].

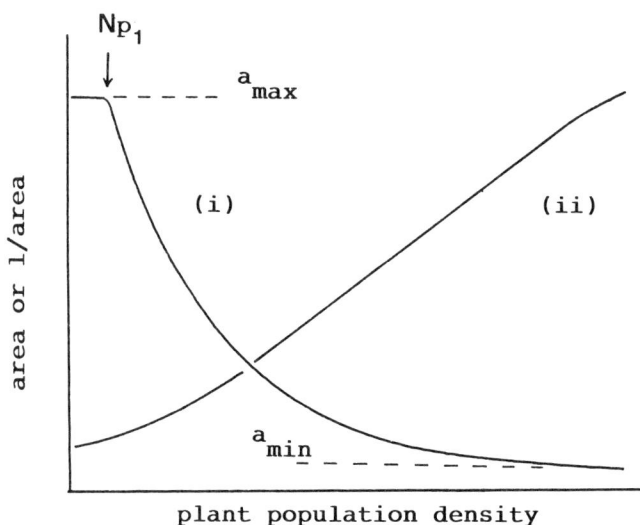

Fig. 2.16. **Representation of the effect of plant population density on leaf area per plant (curve i) and its reciprocal (curve ii).** Text gives explanation.

The intercept of the relation is usually quantitatively similar to the reciprocal of a_{max}, and is smaller for large plants than small ones. If the units of N_p and 1/area are again m^{-2}, the intercept ranges from about 6.0 for a short-season pigeon pea to 0.003 for oil palm[28]. The slope and intercept have significance for the optimum population for yield (as examined in the final chapter).

Influence of N_p on rooting and the R/L ratio in dry environments

The effect of N_p on root length per plant seems, from the little evidence, to be similar to that on leaf area. Both the rate of vertical penetration of roots and the total rate of root extension (unit field area) are much influenced by plant population density in dry conditions. Both rates increase with rise in population, whereas the root length per plant usually decreases as soon as roots of adjacent individuals intermingle.

As an example, Table 2.3 summarizes the responses of groundnut grown on stored water in India at three populations. In wet seasons, stands would normally be established at about the highest of these populations, 23 m^{-2}. In the experiment, the R/L ratio at this population was about four times the value it would have been in a wet season. As population decreased from 23 to 6.6 m^{-2}, the leaf area per plant changed little, but the root length per plant more than doubled. (The reciprocals of both leaf area and root length seemed linearly related to N_p, but the slope was relatively steeper for root length than leaf area.) Consequently, the R/L ratio more than doubled, and at the lowest population was about nine times the value at 23 m^{-2} in moist soil. Pearl millet in Niger[26] responded similarly to a decrease in population, but the effect on R/L was not as great as for groundnut. Possibly the effect is greater for plants, like this groundnut cultivar, whose shoot system is bunched and compact and less able to explore the space above ground than the root system is below ground. (In Chapter 4 this response to population will be shown to affect the rate of transpiration per unit leaf area.)

Table 2.3. Effect of plant population density on root length per unit leaf area (R/L). For groundnut at Hyderabad, India, grown on a drying soil; during expansion, at 76 d after sowing.

Population (m^{-2})	6.6	12.2	22.9
Root length (km)			
per m^{-2} field	3.2	3.7	4.7
per plant	0.48	0.31	0.20
Leaf area (m^2)			
per m^2 field (L)	0.52	0.94	1.66
per plant	0.079	0.078	0.072
R/L (km m^{-2})	6.2	3.9	2.8

Source: Rao *et al.* (1989) (with additional data supplied to author).

Concluding remarks

For most processes considered in this chapter, the effects of temperature are far better understood than the effects of other environmental variables. This is partly a result of the large amount of work on responses to temperature in recent years. Moreover, the principles underlying the responses to temperature, that have been developed from experiments in petri dishes, growth cabinets and glasshouses, seem to apply equally to stands in the field. This is not so of responses to drought and nutrients – which can be very different in stands than in single plants in controlled environments.

The control of duration is far better understood than the control of rate. Though both duration and rate are governed equally by temperature, durations are much less sensitive than rates to other factors, particularly drought, nutrient shortage and, in the case of roots, soil physical properties. Of the various rates, those controlling expansion of the canopy are understood best, though there are still many obstacles to modelling leaf area index of most species in dry, infertile environments (see discussion at the end of Chapter 3). Knowledge of the control of longevity and senescence at the scale of the stand, and of all the processes controlling the dimensions of root system, are still rudimentary. The inaccessibility of root systems contributes to this state, but progress in understanding the appropriate rates might be faster if the canopy and root system were treated as an entity, rather than as separate structures.

References

1   Littleton *et al.*, 1979a, give references to and application of logistic curves in a comprehensive account of the control of leaf growth by temperature (for cowpea) in the field. See also Constable and Rawson, 1980 (cotton).

2   Emergence and soil temperature, Keating and Evenson, 1979 (cassava), Mohamed *et al.*, 1988b (groundnut and pearl millet), Angus *et al.*, 1981 (many species). Maturation temperature and emergence, Mohamed *et al.*, 1985. Supra-optimal temperature, Soman and Peacock, 1985. Figure 2.1 is from J. Garcia-Huidobro, 'Control by temperature and soil water of germination and emergence in pearl millet', Ph.D. thesis, University of Nottingham, 1982.

3   Radicle elongation rate, Blacklow, 1972 (maize), also Gregory's (1987) interpretation of Nishiyama's data for rice. Lamina extension rate, with electronic auxanometers in controlled environment glasshouses, roots in field soil. For pearl millet, Squire and Ong, 1983 (temperature, saturation deficit), Squire, Black and Ong, 1983 (turgor, assimilate), Mohamed *et al.*, 1988c (temperature). For sorghum, Hamdi *et al.*, 1987 (saturation deficit, assimilate). For groundnut, Ong *et al.*, 1985 (saturation deficit, turgor). 'Further reading' (this chapter) gives references to other controlled environment work. For shoot extension rate, see ref. 6. For derivation of base temperature from field data, Stephens and Carr, 1990.

4   Cooper and Law, 1978, Cooper, 1979.

5   Cassava, in Columbia; Connor *et al.*, 1981, and Connor and Cock, 1981 (drought, water potential, R/L ratio), Connor and Palta, 1981 (expansion and leaf water potential), Irikura *et al.*, 1979 (altitude and temperature), Veltkamp, 1985 (photoperiod). In Australia; Aresta and Fukai, 1984 (root length), Fukai *et al.*, 1984 (canopy, shading). And for an example of conservation of nutrient content per unit leaf area, Table 15.6 in the chapter by J. H. Cock, *Cassava*, pp. 529–49 in *The Physiology of Tropical Field Crops* (ref. in General reading, following Introductory remarks).

6 Tea, extension rate in the field, Squire, 1979 (temperature, saturation deficit), Tanton, 1981 (growth curves), 1982a, b (temperature, saturation deficit, photoperiod, assimilate), Carr *et al.*, 1987 (response to soil water), Stephens and Carr, 1990 (temperature). Table 2.2, unpublished data of S. M. O. Obaga, Tea Research Foundation of Kenya, Kericho, Kenya.

7 Examples of response to population, maize, Allison, 1969, pearl millet, Carberry *et al.*, 1985, cassava, Enyi , 1972b, Enyi, 1973, pigeon pea, Akinola and Whiteman, 1974, Tayo, 1982, oil palm, Corley, 1973.

8 Cereals, Gregory and Squire, 1979 (pearl millet), Brown *et al.*, 1987 (barley, in Syria).

9 Extended daylength, earliness, Alagarswamy and Bidinger, 1985, Carberry and Campbell, 1985, Craufurd and Bidinger, 1988.

10 Altitude, Cooper, 1979 (maize, Kenya), Fischer and Palmer, 1984 (maize, Mexico). But see also laboratory experiments and general discussion by Coaldrake and Pearson, 1986 (pearl millet).

11 For example: groundnut, unpublished data of B. Marshall and A. C. Terry, University of Nottingham; summarized in *Microclimatology in Tropical Agriculture* (Introductory remarks).

12 Sheldrake and Narayanan, 1979 (pigeon pea).

13 Monteith and Elston's (1983) analysis of data from Hayashi and Ito.

14 Stirling, Williams, Black and Ong, 1990 (shade), Stirling, Ong and Black, 1989 (drought).

15 Feddes, 1972, gives the analysis, with examples of temperate species.

16 Acevedo *et al.*, 1979 (maize), Turner *et al.*, 1986b (rice); for review and discussion, the article by Hsiao, T. C., Silk, W. K. and Jing, J., pp. 239–66, in *Control of Leaf Growth* (Further reading, this chapter).

17 Turk and Hall, 1980b.

18 Groundnut, Harris *et al.*, 1988, Matthews *et al.*, 1988a, b. Sorghum, Matthews *et al.*, 1990a, b.

19 The saturation deficit is the difference between the saturation vapour pressure at air temperature (Murray, 1967) and the actual vapour pressure of the air, which is usually obtained from wet- and dry-bulb temperatures. It is influenced by both the air temperature and the frequency of rainfall (Monteith, 1986b), so changes much with season and altitude. Examples of daily maxima: high altitude or cool season, moist, 0.5–1.0 kPa; cool season, dry 1.0–1.5 kPa; warm wet season (e.g. mean temperature 25–28°C), 1.5–2 kPa; hot dry season, 4 kPa or larger. See also Chapter 4.

20 Allison, 1969, 1984 (population, maize in Zimbabwe), Muchow, 1988a (nutrients, maize and sorghum in Australia). The data in Figure 2.11 are extracted from Figure 3 in Allison's 1969 paper, and Figure 5 in his 1984 paper, and from Figures 6 and 7 in Muchow (rather than Figures 8 and 9). The conclusions of the analysis are similar if extension rate during the later part of phase II is used instead of the rate from sowing to the end of phase II.

21 Gregory, 1983, 1986 (pearl millet, temperature), Blum and Ritchie, 1984 (sorghum, dry soil).

22 For further rooting profiles in moist soil, see Angus *et al.*, 1983, Mugah, J. O., pp. 80–7, in EAAFJ Special Issue 1984, reference 2, Chapter 4 (maize, California). For genotypic differences in descent, Puckridge and O'Toole, 1981 (rice).

23 For discussion (and references) of the effects of soil physical conditions on root extension, see pages 131–7 in *Russell's Soil Conditions and Plant Growth* (Reference in Introductory remarks).

24 Barley in Syria. Brown *et al.*, 1987 (for R/L ratios cited, R came from Table 5,

and L from Figure 2 in that paper); see also Gregory, Shepherd and Cooper, 1984, for effects of nutrients and water on the distribution of root length throughout the soil profile.

25   Rees, 1986b, implication only, no direct evidence that drought affected root extension via assimilate supply.

26   R/L ratios of pearl millet and groundnut, compare Gregory and Reddy, 1982 (moist) with Azam-Ali, Gregory and Monteith, 1984a (pearl millet, dry, three populations), and Rao *et al.*, 1989 (groundnut, dry, three populations).

27   For example: Brouwer, R. (1983). Functional equilibrium: sense or nonsense? *Netherlands Journal of Agricultural Science* **31**, 335–48.

28   Population and plasticity. As far as can be judged from data in the papers (mainly in graphs), the values of a and b in the regression, $1/\text{area} = a + bN_p$ (units of 1/area and N_p both m^{-2}), are, respectively, 6.4 and 0.17 for pigeon pea (Rowden *et al.*, 1981), 0.82 and 0.11 for rice (Akita, 1982a, b), 0.58 and 0.076 for maize (Allison, 1969), 1.3 and 0.11 for pearl millet (Carberry *et al.*, 1985), and 0.0029 and 0.027 for oil palm (Corley, 1973).

Further reading

The first, fourth and sixth books in the list include much on expansion at levels from the cell to the stand. For general accounts of expansion in the field, see (for canopies) Monteith and Elston, and (for roots) Gregory, and pp. 113–67 (Chapter 4) in *Russell's Soil Conditions and Plant Growth*, 11th Edition (General reading). The last two references in Further reading for Chapter 3, and that by Feldman examine responses at the level of the cell and organ.

Baker, N. R., Davies, W. J. and Ong, C. K. (1985) *Control of Leaf Growth*. Cambridge University Press, Cambridge.

Feldman, L. J. (1984) Regulation of root development. *Annual Review of Plant Physiology*, **35**, 223–42.

Gregory, P. J. (1987) Development and growth of root systems in plant communities. In: *Root Development and Function*. Cambridge University Press, Cambridge, pp. 147–66.

Gregory, P. J., Lake, J. V. and Rose, D. A. (1987) *Root Development and Function*. Cambridge University Press, Cambridge.

Monteith, J. L. and Elston, J. (1983) Performance and productivity of foliage in the field. In: Dale, J. E. and Milthorpe, F. L. (eds), *The Growth and Functioning of Leaves*. Cambridge University Press, Cambridge, pp. 499–518.

Russell, G., Marshall, B. and Jarvis, P. G. (1989) *Plant Canopies: their Growth, Formation and Function*. Cambridge University Press, Cambridge.

Chapter Three

Dry Matter Production by Interception and Conversion of Solar Radiation

Introduction

Previous chapters examined how the environment and the physiological attributes of a plant determine the size and longevity of the leaf canopy, which forms the surface for intercepting solar radiation and exchanging carbon dioxide with the air. For much of a plant's life, there is a net gain of CO_2, leading to production of dry matter, but conditions arise, such as when foliage is senescing or droughted, during which losses are similar to, or greater than, gains, and production ceases.

The rate at which a stand produces dry matter, and the amount produced by the time it is harvested, both depend on many environmental and physiological factors. Tropical species however, can be separated into two groups that differ in the maximum rate at which they can produce dry matter. Plants with the C3 photosynthetic mechanism, such as rice, legumes, root crops and palms, have maximum growth rates (measured over a few weeks) of 35–40 g m^{-2} d^{-1}. Their maximum annual production depends largely on the longevity of the foliage. A stand growing for 100 d produces about 12 t ha^{-1}, whereas one growing throughout a year produces 40 t ha^{-1}. Plants with the C4 photosynthetic pathway, the tropical cereals and grasses, have corresponding maximum rates of 50–55 g m^{-2} d^{-1}. Their maximum annual production increases from 20 t ha^{-1} for species growing for 100 d to 60 to 70 t ha^{-1} for those, such as sugar cane, present all the year[1].

Very few stands grow at these rates and produce such large amounts of dry matter, and C4 stands do not always produce dry matter faster than C3. Among stands examined later in this chapter, the growth rates of the most efficient C4 plants are five times those of the least efficient C3 plants (Fig. 3.1). The C4 stands as a whole produce dry matter more rapidly than the C3, but some stands of both groups produce dry matter at similar rates. Other stands, in dry conditions and on infertile soils, grow much more slowly than many of those in Figure 3.1. In these conditions, the advantage of the C4 pathway usually remains, but in very dry environments, there is sometimes

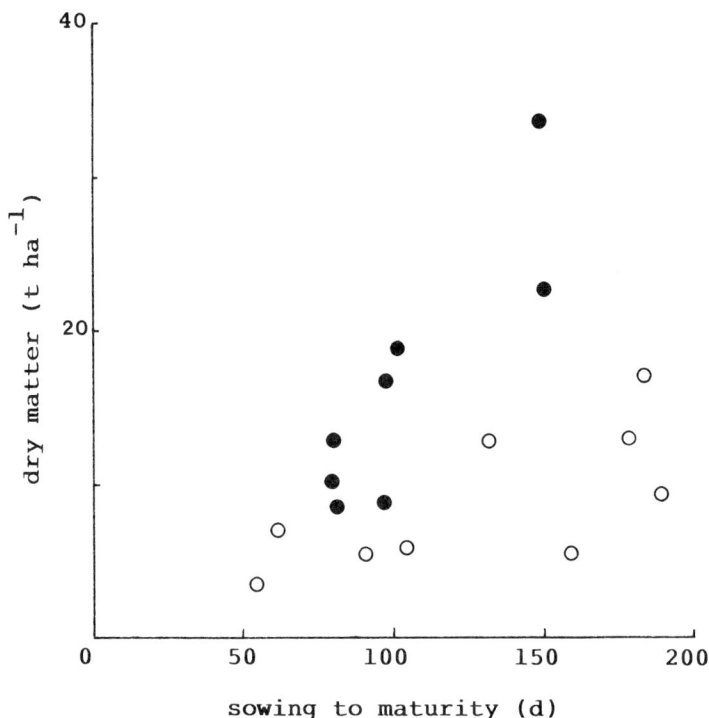

Fig. 3.1. Relation between crop duration and dry matter (above-ground) at harvest. For stands of C4 species (●) maize, sorghum, and pearl millet, and C3 species (○), cowpea, pigeon pea, groundnut, rice and cassava; in moist conditions and with fertilizer.

little difference in growth rate between C4 plants, such as maize and sorghum, and C3, such as cotton and cowpea.

The first step in analysing such differences in dry matter production is to determine how much production depends on the incoming solar radiation and how much on the efficiency with which the stand uses the radiation to produce dry matter. This efficiency is then reduced to a series of factors governed by the physiological attributes of the plant. These attributes can be separated into two broad groups: those, such as the dimensions, configuration and longevity of the canopy, that determine the fraction of incoming energy intercepted and absorbed by foliage; and those that determine rates of photosynthesis and respiration, and thereby the rate of dry matter production per unit of foliage.

The contributions of these two groups of physiological attributes can be distinguished by expressing dry matter as a function of *intercepted* solar radiation[1]. The weight of a stand (W) is therefore related to the cumulative total of intercepted radiation (ΣS_i) by

$$W = \varepsilon_s \Sigma S_i \qquad (3.1)$$

The symbol ε_s is the ratio of dry matter to intercepted radiation, sometimes termed the conversion 'efficiency' for intercepted radiation. When W is expressed in g m^{-2} and ΣS_i in MJ m^{-2}, ε_s is g MJ^{-1} (grammes per megajoule).

The advantage of this analysis is that it compares the performance of stands in terms of two characters (ΣS_i and ε_s). Admittedly, the value of each depends on many processes operating at different rates and periods of time in different parts of the canopy. Further work is necessary to investigate limiting processes in more detail. The methodology with respect to formation and structure of the canopy was described in Chapter 2. Additionally, the techniques of investigating gas exchange – which involve enclosing a plant or part of it in a chamber – can be used in the field to study ε_s[2].

The following account considers, first, general features of ΣS_i and ε_s independently, second, their contribution to potential production and, finally, the restrictions imposed on them by drought and nutrient shortage.

Interception of solar radiation

Intercepted radiation is the difference between solar radiation received at the surface of the canopy, S, and that transmitted at the soil, and therefore includes the fraction of incoming radiation reflected from the canopy. Intercepted radiation is usually determined as the difference between that received at the canopy surface and that transmitted through the canopy, as measured by arrays of solarimeters. The incoming radiation itself varies much throughout the tropics. Seasonal means of total solar radiation (in the wavelength range 0.4–3 μm) range from 12 MJ m^{-2} d^{-1} in cloudy upland regions to more than 24 MJ m^{-2} d^{-1} during cropping seasons in some semi-arid regions. Accordingly, canopies are sometimes best compared, not by intercepted radiation (S_i) itself, but by the fraction S_i/S (here termed fractional interception, or f). For a given canopy, this fraction is little affected by the absolute value of S, so is useful for modelling dry matter production. (Further reading, this chapter, gives reference to reviews of the radiation balance of plant canopies).

Leaf area index and fractional interception

For most canopies in moist conditions, fractional interception (f) may be related to the leaf area index (L) of a canopy by the expression

$$f = 1 - \exp(-kL) \qquad (3.2)$$

where k is an extinction coefficient[3]. The fraction of the solar radiation intercepted by a given leaf area therefore increases as k increases. In practice, k can be determined, from relatively few measurements, as the slope of a linear regression of $\ln(1-f)$ on L.

The value of k depends on whether f is measured in terms of total solar radiation in the wavelength range 0.4–3 μm (f_T), or photosynthetically active radiation in the range 0.4–0.7 μm (f_p). Measurements using tube solarimeters (sensitive to total radiation) and quantum sensors (sensitive to the quantum flux density of photosynthetically active radiation) can be interconverted using the empirical relation

$$\ln(1-f_p) = 1.4 \ln(1-f_T) \tag{3.3}$$

The factor of 1.4 is derived from measurements on several species[4]. The value of $(1-f)$ changes with solar elevation and so should be integrated throughout a day.

The value of k is not independent of the size of a canopy. It sometimes increases as canopies expand to become less clumped and their leaves more randomly oriented; and sometimes decreases if leaves become more vertically oriented as canopies become denser with age, an effect enhanced at higher population density. Nevertheless, k is reasonably stable for a genotype over a wide range of conditions, and differs consistently between some canopies of different architecture (Fig. 3.2). The value of k (total radiation) ranges between 0.3–0.45 for stands whose leaves are mostly held at a steep angle (e.g. most cereals), or are clumped with much overlapping of foliage (e.g. the palms). It increases to about 0.8 in species with leaves held horizontally or well distributed (e.g. groundnut, cassava). Within species containing a range of morphological types, e.g. rice[5], the variation in k can be as great as that between the different species in Figure 3.2. Among other less morphologically variable species (e.g. cassava, oil palm), k differs little between most genotypes.

Seasonal change in fractional interception

The rate at which a stand intercepts solar radiation depends on the rate at which it receives radiation and on f, which is (as shown in the preceding section) governed by attributes of the canopy, mainly leaf area index and the extinction coefficient. The amount of solar radiation intercepted depends also on the duration of the foliage. Therefore canopies should also be described in terms of characteristics such as a seasonal mean value of f (seasonal $\bar{f}$), measured between sowing and harvest, and an annual mean (annual $\bar{f}$).

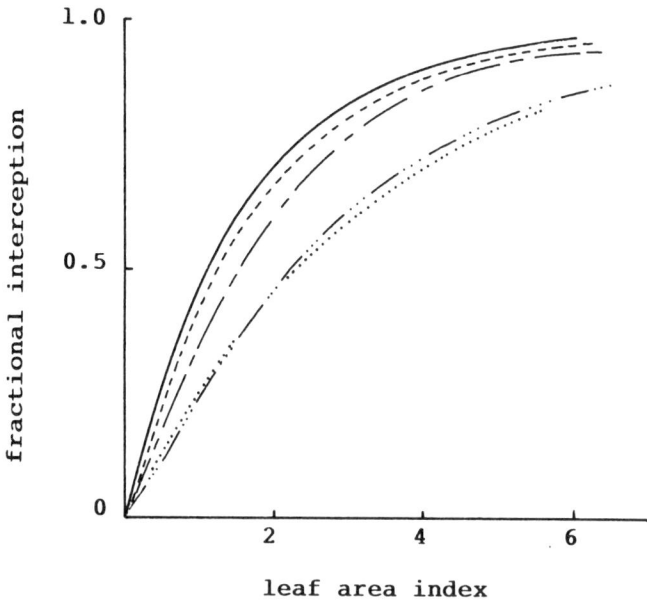

Fig. 3.2. Leaf area index and fractional interception of total solar radiation, as influenced by the extinction coefficient, k.

Species		k	Source
groundnut	———	0.6	Marshall and Willey, 1983
cassava	− − − − −	0.58	Veltkamp, 1985
pigeon pea	—— −−	0.52	Natarajan and Willey, 1980b, Sivakumar and Virmani, 1980
oil palm	— · · ·	0.34	Squire, 1984[12], Breure, 1988a
pearl millet	. . . . . .	0.30	Craufurd and Bidinger, 1988

Effects of growth habit, population and composition of stands

The attributes of a canopy, such as its extension rate, size and longevity, all have some influence on seasonal f̄, as illustrated by a comparison of four species in Figure 3.3. Fractional interception rises more rapidly for a cereal such as sorghum compared with a legume such as groundnut, as a result of a corresponding difference in the rates of leaf development and expansion. Fractional interception also rises more rapidly in both these species than in cassava and sugar cane, which are established at a much lower population density.

Generally, however, differences in f between canopies are smaller than corresponding differences in leaf area index. The asymptotic form of the relation in Figure 3.2 is largely responsible for this; but species with more

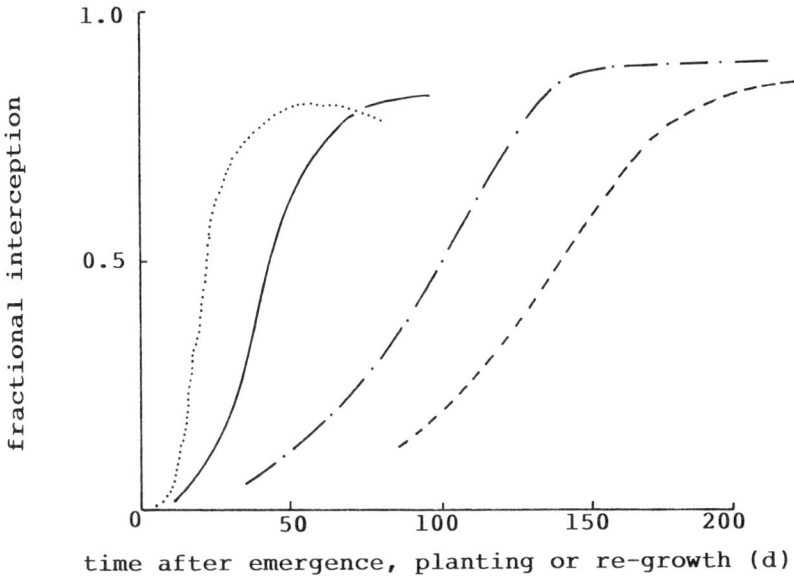

Fig. 3.3. Change of fractional interception of total solar radiation with time. For short-season sorghum (........), groundnut (———), cassava (——— ·) and a ratoon stand of sugar cane (–––). Sources: Natarajan and Willey, 1980b, Reddy and Willey, 1981, Veltkamp, 1985, Batchelor *et al.*, 1985.

slowly expanding canopies commonly have a larger extinction coefficient than those with more rapidly expanding canopies (cf. pearl millet and groundnut). Consequently, most canopies grown in moist, fertile soils differ little in maximum f. The causes of differences in seasonal f̄ lie mainly in the duration of the effective canopy.

The factors that have the greatest influence on this duration are those that control development and longevity of the foliage. The seasonal and annual f̄ are therefore smallest (at about 0.5 and 0.15, respectively) for a short-season cereal or legume, and largest (about 0.9) for a perennial, averaged over its life-cycle. Between these limits, long-season (late) cultivars of species invariably intercept more radiation (Table 3.1), as do the faster expanding C4 cereals compared with legumes among cultivars of a similar duration. Table 3.1 indicates that, in equable conditions, the percentage of solar radiation intercepted during the year increases by 2.4% for each 10 d of growth longer than the 80 d or so required by a short-season cereal or legume to reach maturity. The high value of r^2 indicates the relatively small contribution to annual f̄ by differences between the stands in expansion rate and senescence.

Population density influences the duration of the canopy by affecting the period at the beginning of the season when f is small; it has little influence on

Table 3.1. Comparison of fractional interception of solar radiation for crops of different duration. Seasonal f is the mean fractional interception (total radiation) from sowing to harvest; annual f is the mean for the year (assumptions as in text). Senescence caused some reduction of f for short- and mid-duration cultivars.

Species	Duration (d)	Seasonal f	Annual f	Source
Sorghum	80	0.49	0.11	Natarajan and Willey, 1980b
Groundnut	100	0.53	0.15	Reddy and Willey, 1981
Groundnut	130	0.64	0.23	Bell *et al.*, 1987
Pigeon pea	155	0.51	0.22	Natarajan and Willey, 1980b
Sorghum	150	0.76	0.31	Matthews *et al.*, 1990b
Cassava	300	0.6	0.49	Veltkamp, 1985
Oil palm	(25 years)	–	0.88	Squire, 1984[12]

The linear regression of annual f on duration (t) is $f = 0.0024\,t - 0.099$, $r^2 = 0.92$

the subsequent period when $\bar{f}$ is large. Generally, population has only a moderate effect on seasonal $\bar{f}$. For example, seasonal $\bar{f}$ of pigeon pea in Trinidad increased by a factor of 1.3 (from 0.43 to 0.55) as N_p increased threefold from 20 to 60 m^{-2}. Likewise, the seasonal $\bar{f}$ of groundnut in Australia increased by a similar amount as population density increased almost sevenfold, from 9 to 60 m^{-2}[6].

Mixed cropping of cereals and legumes – where the cereal is usually harvested well before the legume – has little effect on the close relation between annual fractional interception by the canopy and crop duration. The two species together form a canopy a little slower than the cereal grown alone, but when the cereal is removed, f falls to a small value. The legume then continues to expand, and f sometimes rises close to its original value before the cereal was removed. Overall, the mixed crop is present for longer, and so intercepts more solar radiation, than the cereal, but the mixed crop and legume intercept a similar amount of radiation. For instance, values of mean f (for the period between sowing and harvest of the legume) for cereal, legume and mixed crop were respectively 0.39, 0.53 and 0.45 for pearl millet and groundnut, and 0.36, 0.51 and 0.53 for sorghum and pigeon pea[7].

Temperature and determinacy

Temperature has a stronger effect than either population or mixed cropping on the duration of stands, and thereby on seasonal $\bar{f}$. It was shown in Chapter 2 that, when water and nutrients were not limiting, the formation of a canopy can sometimes be treated as a composite process in which leaf area index depends on the passage of thermal time. Provided temperature has little effect on the extinction coefficient (and there is little evidence that it

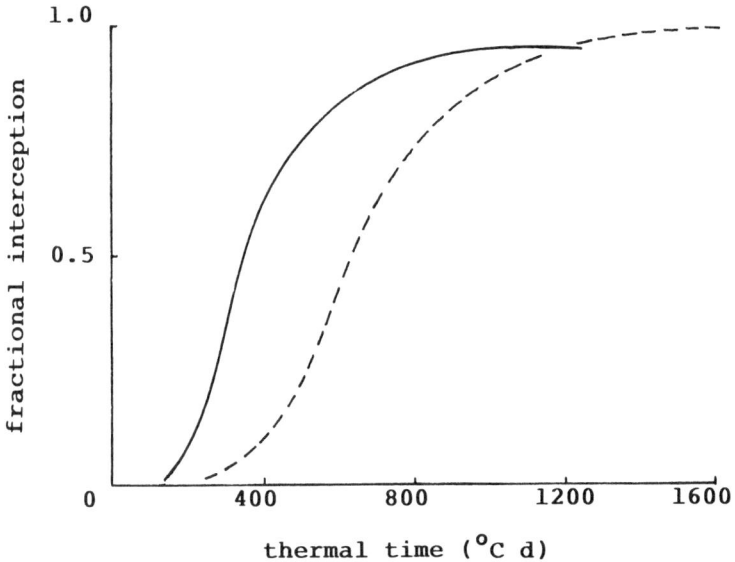

Fig. 3.4. Fractional interception (f) as a function of thermal time from sowing. From measurements in controlled environment glasshouses for (———) pearl millet and (– – –) groundnut. See Reference 8.

has), this analysis can be extended to show that the value of f at any time after sowing is also a function of thermal time (Fig 3.4).

Accordingly, the effect of temperature on seasonal $\bar{f}$ depends much on whether a stand is determinate or indeterminate (as considered for leaf area index in Chapter 2). The effect of this difference is now illustrated using attributes of groundnut, a species which can sometimes be treated as determinate, and sometimes indeterminate. In both modes, seasonal $\bar{f}$ can be computed from a relation between f and time, such as a logistic curve[8]. The derivation of seasonal $\bar{f}$ is simplified in the following examples by considering interception in two phases. The first is canopy formation, i.e. from sowing to an arbitrary maximum of $f = 0.9$ (f_m), in which mean f in groundnut (and incidentally in pearl millet also) is about 0.35 (f_c). The second phase is that after f_m is first reached, when f remains at f_m. Even if leaf area index increases throughout the second phase, the absolute change in f will be very small.

When the canopy is *indeterminate*, only the period t_c from sowing to f_m is governed by a thermal duration ($\hat{\Theta}_c$) and at any temperature between T_b and T_o is given by $\{\hat{\Theta}_c/(T - T_b)\}$. The total period between sowing and harvest (t_t) depends on other factors, and the period at f_m is the difference between t_t and t_c. Therefore, seasonal mean $\bar{f}$ is given by

$$\bar{f} = \{f_c t_c\} + \{f_m(t_t - t_c)\}/t_t \qquad (3.4)$$

For example, if the time from sowing to harvest of a crop (t_t) is 120 d, Θ_c is 1000°Cd, and T_b is 10°C, then the period, t_c, would decrease from 100 d at 20°C, to 50 d at 30°C; and the period at f_m would increase from 20 d at 20°C to 70 d at 30°C. Accordingly, when $f_c = 0.35$ and $f_m = 0.9$, seasonal $\bar{f}$ would *increase* from 0.44 at 20°C, to 0.67 at 30°C, and annual $\bar{f}$ from 0.14 to 0.22, respectively.

When the canopy is *determinate*, the period from sowing to f_m is governed by a thermal duration ($\hat{\Theta}_c$) as in the previous example, but the period at f_m is also governed by a thermal duration ($\hat{\Theta}_m$). Therefore seasonal mean fractional interception is given by

$$\bar{f} = \{(f_c\hat{\Theta}_c)+(f_m\hat{\Theta}_m)\}/(\hat{\Theta}_c+\hat{\Theta}_m) \tag{3.5}$$

and is independent of temperature between the base and the optimum temperatures for development (assuming base temperatures for all processes are similar). Suppose the time-to-harvest is governed by a thermal duration of 2000°Cd, but that all other attributes are as in the previous example. The period from sowing to f_m would still be 100 d at 20°C and 50 d at 30°C, but the time spent at f_m would decrease from 100 d at 20°C to 50 d at 30°C. Although f averaged from sowing to harvest would be 0.62 irrespective of temperature, annual mean $\bar{f}$ would *decrease* from 0.34 at 20°C to 0.17 at 30°C – a response opposite in direction to that for the indeterminate case.

The relation for the determinate case defines seasonal $\bar{f}$ for genotypes whose durations depend on only temperature, and must be modified if duration is sensitive to photoperiod. The term $\hat{\Theta}_c$ in equation 3.5 will not, in general, be much affected, since even in short days most species produce enough leaves in good growing conditions to intercept 90% of the available solar radiation; and although more leaves are produced at longer photoperiods, the extra leaf area intercepts little extra solar energy. The main effect of photoperiod will be on the duration of the period at f_m, which for short-day plants increases with rise in photoperiod.

Conversion of intercepted radiation

The ratio (ε_s) of dry matter produced to solar radiation intercepted can be based on a single large sample of dry matter at harvest, or derived from a linear regression of dry mass on intercepted radiation for several samples during a defined period of growth. Roots are not included in most estimates of ε_s. Their mass is a relatively small proportion of the total (10–20%) for annual herbs growing in moist conditions, but can be a very substantial proportion (e.g. 50%) for herbs in dry conditions and for perennial crops generally (see Chapter 5).

Comparisons of stands by ε_s are valid only when the plant material in each stand has a similar energy equivalent. The nutrients in plant tissue are included in the estimate, but differences in the nutrient content of different stands would hardly affect the absolute values of ε_s. More important is the presence of energy-rich substances that accumulate during reproductive growth in many species. The use of dry mass as the basis of ε_s will only slightly underestimate the true efficiency of radiation use in a cereal, but by as much as 30% in a productive oil crop. For groups such as legumes and palms, the dry mass of oil can be converted to an equivalent of carbohydrate or assimilate, in order to derive an equivalent stand dry weight, W^*, and the corresponding conversion ratio, ε_s^*[9].

In the following analysis, the conversion ratio is calculated on the basis of total solar radiation, with the value in photosynthetically active radiation (PAR) sometimes given for reference[4].

Change during the life-cycle

Some of the carbon that plants fix in photosynthesis is expended in the respiration associated with maintaining existing tissue – the 'maintenance respiration' – and producing new tissue – the 'growth respiration'[10]. There is little information on the magnitude of these two types of respiration in stands in the field, but there has developed a notion, particularly in the early

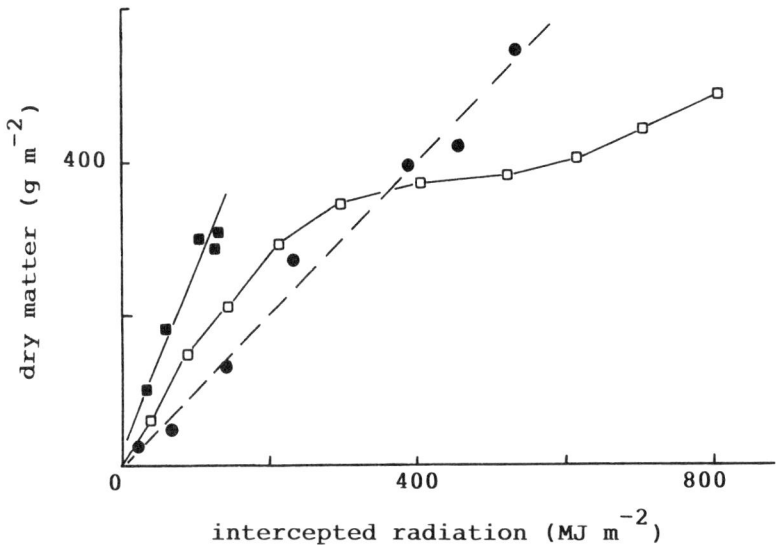

Fig. 3.5. Dry matter and intercepted radiation for stands of groundnut. At Hyderabad, India, in wet season: ■, shaded, slope is 2.4 g MJ^{-1}, ●, unshaded in same experiment, 1.0 g MJ^{-1}, both from Stirling, Williams, Black and Ong, 1990, □, unshaded, different year, adapted from Marshall and Willey, 1983.

literature of crop physiology, that the maintenance respiration is proportional to the mass of living tissue. The implication is that ε_s should decrease as the dry mass of a stand increases.

The relation between ε_s and mass is difficult to determine in annual crops, in which age has independent effects on photosynthesis and respiration. The dry matter accumulated by annuals commonly increases linearly with the amount of solar radiation intercepted, at least for a substantial part of the life-cycle. In determinate canopies, linearity is usually maintained throughout the period when new leaf is being produced, but can remain for much longer in indeterminate canopies that produce new leaf throughout the period of fruiting (e.g. pigeon pea, groundnut). This conservatism of ε_s implies that (a) the response of photosynthesis to solar radiation varies little between the parts of the canopy contributing most to dry matter production; and (b) the loss of CO_2 is a conservative fraction of the gain.

For many canopies, however, ε_s decreases after flowering, though stands differ much in the extent of the decrease. Large differences have been found even between stands of the same genotype grown in wet seasons and given fertilizer (Fig. 3.5). The physiological basis of the decrease is difficult to determine. It might indicate an increase in the respiration rate, but it is more likely the result of a decrease in the photosynthesis rate as leaves age[11], or as nutrients are exported from them to fruiting structures (see below). Commonly, the amount of solar radiation intercepted after the start of fruiting is similar to that before, so the reduction of ε_s during fruiting greatly reduces total productivity.

Relations with plant mass in perennials

Most of the evidence suggests that ε_s changes little with increase in the standing dry mass of tropical perennials. In studies of oil palm in Malaysia, Nigeria and Papua New Guinea[12], ε_s was sometimes found to change when the canopy was expanding (though it was unclear why it changed), but then changed little while the dry mass of the stands increased from 20 to 90 t ha^{-1} (Fig. 3.6). The increase in mass was largely a result of growth of the trunk. Modelling the carbon balance of the palms indicated that respiration per unit mass of palm tissue as a whole should decrease as plants grow larger. The implication is that respiratory losses are mainly confined to the apex of the trunk, and to the leaves and the fruits, whose combined mass changes little after the canopy closes: the bulk of the trunk tissue must be respiring very slowly.

Relations with mass in population experiments

Effects of age and mass can sometimes be separated by analysing the response of ε_s to plant population density. As population increases, the tendency is for ε_s, based on above-ground dry matter, to be stable (as in groundnut), or even to increase[13]. For example, ε_s almost doubled both for pigeon pea in Trinidad, as density was raised from 20 to 60 plants m^{-2}, and for oil palm in Malaysia as density was raised from 100 to almost 400 ha^{-1}. The physiological mechanisms controlling this response are unknown. A rise in planting density could have several effects. It generally reduces the fraction of a plant's store of dry matter that is allocated to fruiting structures (see Chapter 5). More assimilate is respired during the making of fruits than vegetative tissue, so the effect might reduce the losses of CO_2 from the stand. Alternatively, a rise in density might cause leaves to be held more vertically, in which case less light is intercepted at the canopy surface, and more penetrates deeper, thereby increasing the mean photosynthetic efficiency of the foliage (see p. 84 for explanation).

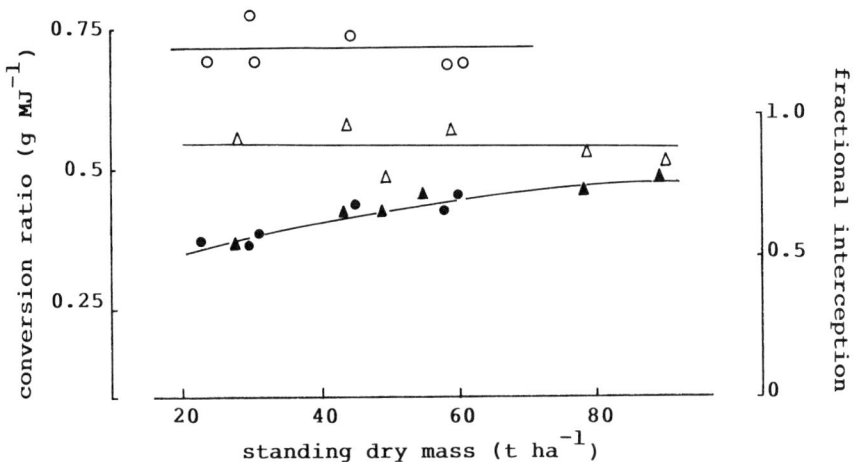

Fig. 3.6. The conversion ratio (ε_s) and fractional interception (f) for total solar radiation in relation to the dry mass of oil palm stands. From the destructive measurements of Rees and Tinker (1963) in Nigeria ($\triangle$, ε_s; $\blacktriangle$, f), and Corley, Gray and Ng (1971) in Malaysia ($\circ$, ε_s; $\bullet$, f), using a common extinction coefficient[12].

Whatever the basis of the response, it results in the maximum growth rate of a genotype commonly being achieved at the largest leaf area index it can maintain. Admittedly, when L is maximum, canopies might produce the least dry matter per unit leaf area, and so have the lowest net assimilation rate (mass of dry matter per unit leaf area per unit time); but this happens because of the asymptotic relation between fractional interception and leaf area index. There is little indication in the field of an optimum L, or standing dry mass, above which ε_s or dry matter production decreases.

Differences within and between species

During vegetative growth, the more efficient tropical cereals with the C4 photosynthetic pathway achieve the largest values of ε_s – about 2.5 g MJ^{-1} total radiation (or 4.2 g MJ^{-1} PAR). Legumes and other crops with the C3 pathway achieve a maximum of about 1.4 g MJ^{-1} (2.5 MJ^{-1} PAR) when exposed to full sunlight as sole crops. Perennials achieve a maximum of about 0.9 g MJ^{-1} (1.6 g MJ^{-1} PAR), averaged over 6–12 months. The effect of the decline in ε_s after flowering in annuals is such that the long-term mean ε_s differs little between the main groups. For example, among C3 species, the long-term value of ε_s for stands with plenty of water and nutrients can be similar for groundnut, chickpea, cassava and the palms (Table 3.2).

Table 3.2. Comparison of dry matter (above-ground) per unit intercepted radiation, ε_s (units: g MJ^{-1}), for stands supplied with fertilizer, in wet seasons: from sowing to flowering (fl.), and to the time of maximum dry weight (max. wt.).

Species	e (total radiation)		e (PAR)		Source
	to fl.	to max. wt.	to fl.	to max. wt.	
Pearl millet*	2.4	1.6	–	–	Reddy and Willey, 1981
Pearl millet*	–	–	4.1	2.9	Marshall and Willey, 1983
Pearl millet	–	2.0	–	–	Craufurd and Bidinger, 1988
Maize	–	1.4	–	–	Natarajan and Willey, 1980b
Maize	1.6	1.3	–	–	Muchow and Davis, 1988
Sorghum	1.3	1.2	–	–	Muchow and Davis, 1988
Groundnut*	1.4	0.8	–	–	Reddy and Willey, 1981
Groundnut*	–	–	2.5	1.2	Marshall and Willey, 1983
Groundnut	–	–	2.8	2.2	Bell et al., 1987
Pigeon pea	–	0.54	–	–	Natarajan and Willey, 1980b
Pigeon pea	0.83	0.72	–	–	Hughes and Keatinge, 1983
Chickpea	0.78	–	–	–	Hughes et al., 1987
Cassava		0.9		1.5	Veltkamp, 1985
Oil palm		0.9		1.6	Reference 12, this chapter
Rubber, cocoa, coconut		–		1.0 to 1.5	Corley, 1986

*Measurements in the same field.

Within a group, ε_s sometimes differs much between stands that seem to be growing in favourable conditions. For example, in Table 3.2, ε_s ranged from 1.2–2.0 g MJ^{-1} for the C4 cereals. It was smaller for sorghum and maize than for pearl millet, though such differences between these C4 species are not found more generally. The differences between these examples were probably caused by unidentified environmental or genotypic factors. The photosynthetic efficiency, even of C4 species, is reduced at very high solar irradiance[14]. The maize and one of the sorghum stands received much more radiation per day than the millet. The value of ε_s is also affected by atmospheric dryness and leaf nutrient content (examined later), both of which might have differed between the sites.

There have been few systematic comparisons of ε_s among genotypes grown at the same time and in the same place. At most there seems to be only moderate variation of ε_s within a species. For example, among four genotypes each of sorghum and groundnut grown at Hyderabad, India, ε_s varied over a range of about 30% of the mean for the four, whereas there were no differences in ε_s detected between several cassava genotypes in Columbia[15].

Conversion ratios of mixed stands

Mixed canopies of a C4 cereal, such as pearl millet, maize or sorghum, and a C3 legume, such as groundnut or pigeon pea, have a conversion ratio, ε_s, similar to or slightly greater than that of the cereal alone, and much greater than that of the legume[16]. Detailed measurements of light interception and growth have been made for both components in several intercrops. Generally, ε_s of the cereal is little influenced by the proximity of the other crop. In contrast, ε_s of the shorter C3 is larger while shaded by the cereal than when grown alone. The intercropped stand therefore uses solar radiation with an efficiency comparable to that of a C4 species.

For example, ε_s for groundnut during vegetative growth was 1.3 g MJ^{-1} (total radiation) when grown alone, but 2 g MJ^{-1} when beneath pearl millet – almost as efficient as the millet itself. Experiments in India, in which artificial shade was substituted for the millet plants, suggested the enhancement of ε_s was caused by shading rather than some other effect of the taller millet[16]. A similar effect of shade on ε_s is implied by work on stands of another C3 crop, cassava[16]. The physiological cause of the enhancement of ε_s has not been systematically investigated, but probably has its basis in the shape of the photosynthesis/light response of C3 species, which results in the light being used more efficiently at low than at high 'intensities'. An improved water status under shade might also contribute, though ε_s is enhanced even in moist conditions. It is not known if, more generally, the nitrogen fixed by the legume increases the efficiency of the cereal.

The effect of temperature

There have been many laboratory measurements of the effect of temperature on the photosynthesis of leaves, but few corresponding studies in the field, where it is difficult to control leaf temperature adequately and to break the coupling between temperature and saturation deficit. For example, a set of cowpea measurements in Nigeria[17] indicated an optimum of 25–30°C, above and below which the maximum photosynthesis rate declined only slightly; between 20–40°C, P_m was mostly between 1–1.5 mg CO_2 m^{-2} s^{-1}. However, it was uncertain the decline above 30°C was caused by temperature *per se*, since the saturation deficit also rose, possibly to a value where it might have restricted photosynthesis (see p. 92). Nevertheless, this study showed temperature had only a small effect on photosynthesis rate.

There is even less information on the effect of temperature on ε_s for stands. In simulated field conditions (controlled environment glasshouses), pearl millet maintained values of ε_s close to the maximum in Table 3.2 for several months, over a wide range of mean temperature. The response was somewhat flatter than might be expected on the basis of typical photosynthesis responses of C4 species, but the photosynthetic system might have adapted so that the optimum temperature for photosynthesis was similar to that in which stands were grown[18]. The shape of the response might have been untypical, since the solar irradiance in the glasshouses was about half that outdoors during monsoon seasons in the tropics. Nevertheless, the weak response is consistent with the apparent effects of temperature on rate of dry matter production in the study of maize and altitude described in Chapter 2. Once the canopies had formed, the maize in Kenya produced dry matter at a similar rate at the three altitudes (allowing for differences in solar radiation).

Synthesis: control of production in moist environments

Dry matter production of stands growing in moist, fertile environments is examined by the analysis described in this chapter. The maximum rates of dry matter production can be achieved only under clear tropical skies (e.g. $S > 24$ MJ m^{-2} d^{-1}) by stands for which f is at least 0.95, and ε_s is the largest value in Table 3.2 appropriate for the group (C4 or C3). The values of maximum annual production are not obtained in climates of continuously high solar radiation, because these climates have a dry period during at least part of the year. For example, the maximum productivity of a C3 species (oil palm) of a little more than 40 t ha^{-1} (carbohydrate equivalent), was achieved in a cloudy climate where S was 16 MJ m^{-1} d^{-1} (total radiation), f about 0.9, and ε_s about 0.8 g MJ^{-1} (total radiation).

The main factors, other than solar radiation, that cause differences such as those between the stands in Figure 3.1 are plant population density, the composition of the stand and temperature. All these affect the three main attributes of a stand in different ways; these attributes being its leaf area, its conversion ratio for solar radiation and its duration.

Population density and stand composition

Neither the plant population nor the composition of a stand has much affect on the timing of development. The population density has a moderate effect on the conversion of intercepted radiation to dry matter, but its influence on production is mainly through leaf area index. Production therefore increases as population rises, and effectively reaches a plateau when further increase in population results in only slightly more intercepted radiation. There is little evidence that stands in the field reach an optimum population density above which production decreases.

The effects of composition are complicated, but the mixed stands that have been examined produced more dry matter than their components alone. The cereal/legume intercrops, referred to earlier in this chapter, intercepted more solar energy than the cereal alone, and about the same amount as the legume. When both cereal and legume were growing together, the intercrop was more efficient than either crop on their own. Even including the period near the end of the season when the legume of the intercrop was growing alone, ε_s for the intercrop was still as much as 80–90% that of the cereal. Consequently, the mixed crops produced more dry matter than either crop grown singly. They intercepted more solar radiation than the cereal alone, and used it more efficiently than the legume. In three cereal/legume mixed crops, dry matter production was 1.1–1.9 times that of the cereal, and 2–2.5 that of the legume. In effect, the mixed crops grew for the duration of the legume at a rate more nearly that of a C4 than a C3. It is not known whether mixed crops can produce dry matter more rapidly than the maximum values for single species stands given at the beginning of this chapter. The experiments have been undertaken in monsoon environments where production is limited by the solar radiation.

Effects of temperature and determinacy

The maximum production of different genotypes within each of the C4 or C3 groups depends mainly on mean fractional interception between sowing and harvest. This is the seasonal $\bar{f}$ in the analysis on p. 78, which is influenced by the determinacy of the genotype, and the control by temperature of the duration of the canopy.

The fundamental basis of the response to temperature, in both determinate and indeterminate genotypes, lies in the relations examined in earlier chapters and on p. 78 between l/duration for a process and temperature above a base. Figure 3.7 shows the results of applying very simple models of these relations for two types of stand: one consisting of a determinate C4 cereal, and the other an indeterminate C3 legume. Values of the base temperature and the thermal durations, and of ε_s at different temperatures,

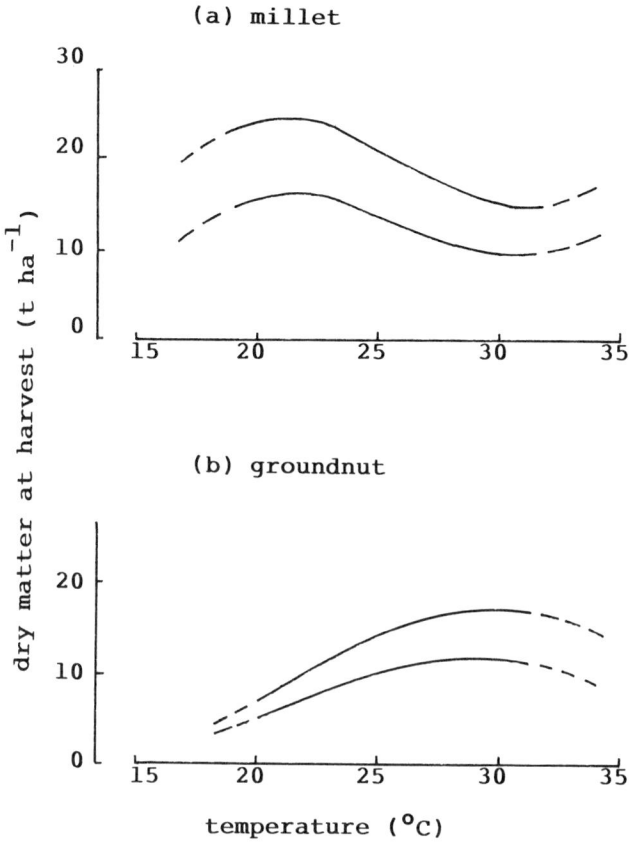

Fig. 3.7. **Modelled response of dry weight at harvest to mean temperature.** For (a) pearl millet hybrid BK 560 and (b) groundnut cv. Kadiri 3. In each of parts (a) and (b), upper lines show the result when the conversion coefficient is constant throughout growth, lower lines when it decreases by 50% after canopy formation. The dashed lines indicate response not well defined. Incoming solar radiation is 15 MJ m^{-2}.

were obtained in controlled environment glasshouses using stands with their roots in field soil. Responses below 20°C and above 30°C are indicated by dashed lines on Figure 3.7 because there is little information about them.

Most dry matter is produced when the product of ε_s and intercepted radiation is greatest. For the determinate genotype, this occurs at quite a low mean temperature, 20–22°C. As temperature increases above this, weight declines (by about 7% per °C) because ε_s changes little while the duration of the canopy decreases, and with it the amount of intercepted radiation. Below 20°C, weight decreases since ε_s is suppressed relatively more than the duration of the canopy is increased. Above 30°C, the direction of the response is uncertain. The positive effect of temperature on duration would tend to increase weight, but the negative effect on survival would tend to reduce it. Additionally, weight would be reduced if high temperature or dry air suppresses ε_s. In contrast, stands of the indeterminate genotype are heaviest at about 30°C. They produce less dry matter as temperature falls (about 11% per °C), mainly because of a reduction in the time during which the complete canopy is present. Above 30°C, the duration of the canopy probably decreases, and perhaps also ε_s, to reduce production considerably.

At about 30°C, the cereal and the legume produce similar amounts of dry matter; the cereal intercepts less radiation than the legume because the canopy exists for a shorter time (a difference between the genotypes rather than the groups generally), but uses the radiation much more efficiently (a difference between the groups). As mean temperature falls, the cereal intercepts more radiation, the legume less, such that at 20°C, the cereal produces four times as much dry matter as the legume.

The modelled responses also show the large fall in production that occurs when senescence reduces ε_s after flowering. The upper curves in both parts (a) and (b) of Figure 3.7 are for stands for which ε_s remains constant from sowing to harvest, and at the maximum respective values in Table 3.2. The lower curves represent stands for which ε_s decreases by 50% when reproductive growth begins. In the determinate genotype, temperatures has little effect on the duration of the senescent canopy as a fraction of the whole period of growth, and therefore little effect on the dry matter of the senescent stand as a fraction of the non-senescent. In the indeterminate genotype, however, a rise in temperature (between 20–30°C) increases the duration of the (post-flowering) senescent canopy as a fraction of the stand's life; production by a senescent canopy is therefore reduced relatively more as temperature rises.

A final point: the models show both general and specific features. The direction of the response in parts (a) and (b) of Figure 3.7 is representative of genotypes whose total duration is determinate and indeterminate, respectively. However, the optimum temperature for dry matter production would be specific to the genotype whose attributes are used in the model. Other genotypes will not necessarily have the same optimum as one of these. For example, a rice genotype grown at three altitudes in Thailand[19] had an optimum of about 20°C, which was similar to the cereal shown in Figure 3.7, but maize in western Kenya had an optimum of 15°C (or even lower). For the more or less indeterminate groundnut in Zimbabwe referred to earlier, dry matter production increased with rise in temperature from 17°C to 23°C – the same direction as the response in Figure 3.7b – but the optimum for the genotype was probably well above the temperature at the warmest site.

Restrictions by drought and nutrient deficiency

The effects of temperature and solar radiation might account for many of the differences in dry matter production among each of the groups shown in Figure 3.1. However, in much of the tropics, drought and poor soil constrain production to values much lower than in Figure 3.1. Cereals in dryland subsistence agriculture commonly produce ten times less than the maximum for the genotype (even in a 'good' year); and plantation agriculture is still practiced when dry matter production is as little as one third to one quarter of the corresponding maximum.

Much is known of how the limiting factors reduce leaf area and photosythesis of parts of plants in controlled conditions, but relatively little of how they affect the interception and conversion of solar energy by stands. The responses of f and ε_s have been difficult to define in the field, where some of the environmental factors that govern them are usually closely coupled (e.g. temperature and atmospheric dryness). The main constraints imposed by drought and nutrients are examined here. Other factors, such as salinity and waterlogging, can be locally dominant, but very little is known of their systematic effects on f and ε_s.

Effects of drought

In most of the experiments mentioned here, the effects of drought were examined in soils which were naturally fertile or which were given what were considered to be adequate amounts of nutrients for the growth of crops. One of the effects of drought, however, is to make nutrients in the soil less available to the plants[20]. In most of the experiments, this effect probably complicates the effects of the water potential or turgidity of plant tissue.

On seasonal interception

Drought always reduces seasonal mean f from the maximum value determined by its response to temperature and photoperiod. Droughted canopies expand more slowly, are smaller and lose leaf area more rapidly. The extent of the reduction in seasonal f caused by these responses is shown for several stands of pearl millet and groundnut grown in a range of conditions (Figs 3.8a, 3.9). For stands growing solely on a store of water in the soil, f was reduced much less during canopy formation, when the roots were expanding into moist soil, than after, when much of the water available to the roots had been used. The effects of dry air are difficult to separate from the effects of soil moisture. Together, these factors reduced f for millet on a drying soil and in very dry air (D = 4 kPa) to 40% of the value in moist monsoonal weather (D = 1 kPa).

(a)

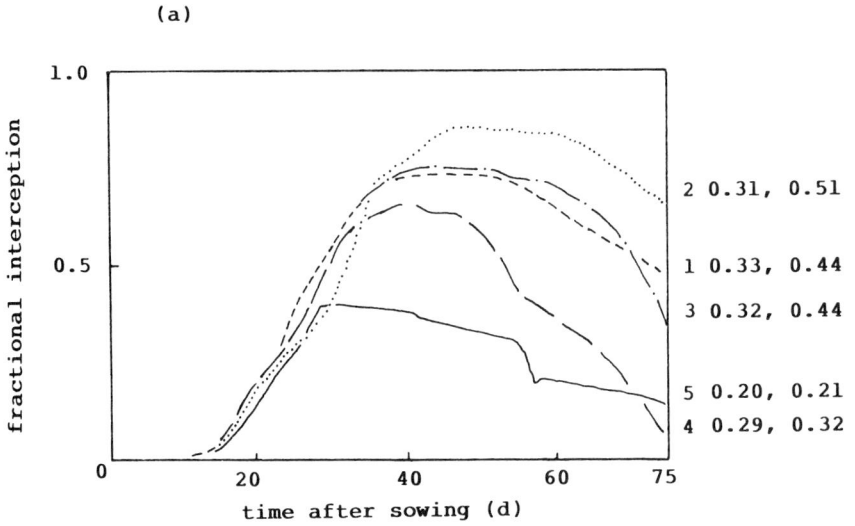

(b)

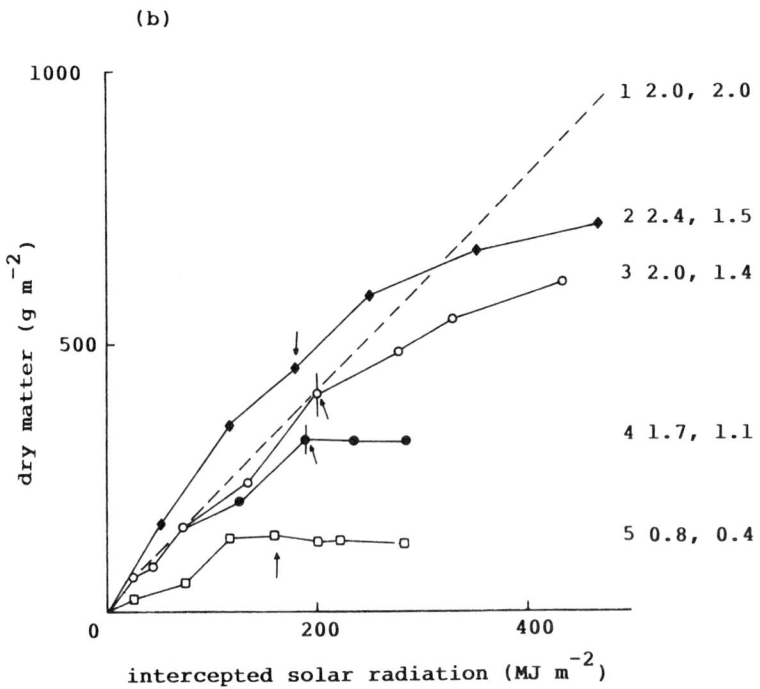

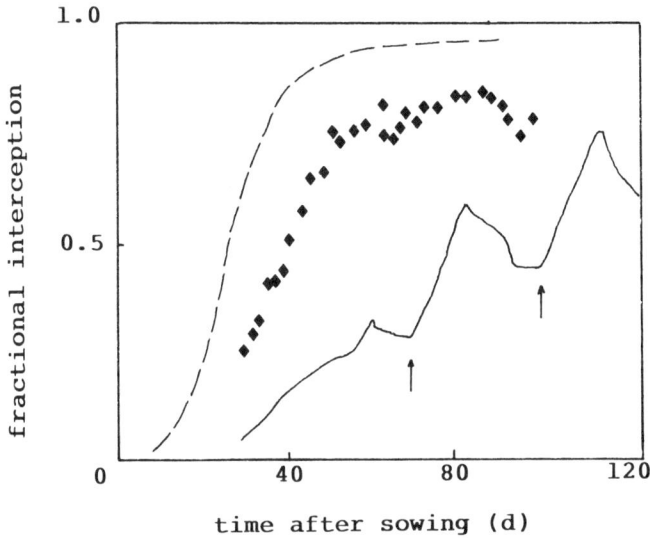

Fig. 3.9. Fractional interception of total solar radiation for stands of groundnut. Mean air temperatures were 25–30°C. Other conditions including daily maximum saturation deficit (D), were: ---, glasshouses, moist soil, D = 1.5 kPa; ♦, Hyderabad, rainy season, D = 1.5–2.0 kPa (Reddy and Willey, 1981); ——, Hyderabad, dry season, with occasional irrigation (arrows), D = 3.0 kPa (Matthews *et al.*, 1988b).

In the groundnut, drought-induced leaf movements (Fig. 3.9), which reduced the extinction coefficient, and so the amount of light the canopy intercepted with a given leaf area[21]. Canopies such as this, whose leaves can roll or move, suffer less leaf-scorch during a drought, though the reason for this is not always clear, since the leaf movements sometimes have little effect on leaf temperature. (The implications of leaf movements for survival, and subsequent growth and yield, were considered in Chapter 2.)

Fig. 3.8. Comparisons of (a) fractional interception of total solar radiation and (b) the conversion ratio for stands of pearl millet. Mean air temperature was 25–30°C. Other conditions, including daily maximum saturation deficit (D) during canopy formation, were:
1 monsoon, Hyderabad, India, mean of several treatments; Craufurd and Bidinger, 1988.
2 monsoon, Hyderabad, D = 1.5 to 2.0 kPa; Reddy and Willey, 1981.
3 dry season, Hyderabad, irrigated, D = 3 kPa; Squire *et al.*, 1984.
4 as for (3), drying soil.
5 dry season, Niamey, Niger, drying soil, D = 4.0 kPa; Azam-Ali *et al.*, 1984b.
In (a) numbers after each curve give mean f between sowing and anthesis at about 45 d (first number) and 75 d (second number); in (b) numbers give mean e from sowing to arrowed sample (generally about 45 d), and from sowing to harvest (about 75 d).

On the conversion ratio

Drought also reduced the conversion ratio, ε_s, of the stands of pearl millet as shown in Figure 3.8. Again, the effects of dry air and dry soil are difficult to distinguish, but appear independent and additive. During canopy formation, ε_s in moist soil and humid air (D = 1.5 kPa) was more than twice the value in drying soil and very dry air (D = 4 kPa). After flowering, which occurred a few days later than the end of canopy extension, the two stands on stored water (parts d and e on Figure 3.8) produced no more dry matter, but retained some green leaf and stem tissue. Consequently, ε_s measured from sowing to harvest for these stands was small – about a half and a fifth, respectively, of the stand with the largest value of ε_s.

Table 3.3. Typical variation in the solar radiation conversion ratio (ε_s) in environments of different dryness, for C4 (pearl millet, sorghum) and C3 (cowpea, groundnut) species.

Soil water	Saturation deficit (daily max. kPa)	ε_s (g MJ^{-1}, total solar)	
		C4	C3
Moist	< 1.5	2.2–2.5	1.0–1.3
Moist	2–3	1.8–2.1	0.8–1.2
Drying	2–3	1.4–1.7	0.5–0.7
Drying	3–5	0.3–0.8	0.2–0.5

Source: various stands in India and Africa, described elsewhere in this chapter.

In this and most other studies, there is insufficient information for ε_s to be expressed quantitatively in relation to an index of dryness, but Table 3.3 gives an indication of the value of ε_s typically found in environments of different aridity. In the driest environments that can support agriculture, there is sometimes little difference in the conversion ratio of a C4 cereal such as sorghum and a C3 such as cowpea[22].

The response of ε_s to drought, based on above-ground dry matter, is sometimes the result of an increase in the fraction of the total dry matter allocated to roots[23]. Generally, however, drought affects ε_s by reducing the rate of photosynthesis[24]. Dry air, independent of the state of water in the soil, contributes to this effect by reducing the leaf (stomatal) conductance, sometimes without affecting the leaf water potential. More generally, both water potential and photosynthesis rate are smaller for stands in dry conditions. Sometimes the photosynthesis rate and the conversion ratio are clearly related to the leaf water potential (as in Fig. 3.10 for groundnut in controlled environment glasshouses), but more commonly in the field these

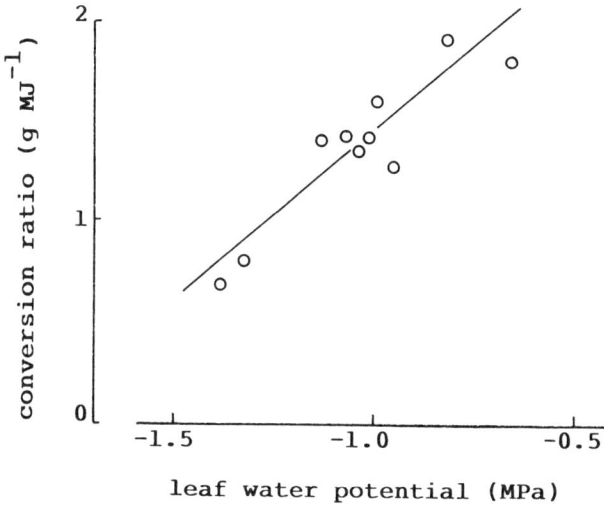

Fig. 3.10. Leaf water potential and the conversion ratio, for groundnut in controlled environment glasshouses. Water potential is a mean between 1000 and 1600 GMT (Ong, Simmonds and Matthews, 1987).

attributes are usually not tightly coupled to the water potential or turgor potential (see later discussion below and Chapter 4).

Relative effects on interception and conversion

The preceding discussion and that in Chapter 2 on leaf area, indicates that for herbs at least, dry conditions might affect leaf area more than photosynthesis. Because of the asymptotic relation between leaf area and fractional interception and because many herbs can move their leaves in order to intercept less radiation, the corresponding reduction in f is commonly more or less proportionately similar to the concomitant reduction in ε_s. This was the case for stands of the short-season pearl millet in Figure 3.8. During canopy expansion, they responded to drought by reducing f and ε_s about equally. After flowering they responded, by reducing ε_s more than f.

Whether dryland annuals reduce either f or ε_s to the greatest extent in response to drought depends not just on the genotype but on environmental conditions, particularly the timing of the drought in relation to the phase of canopy expansion. For example, in a series of experiments with sorghum in Botswana[22], seasonal interception was reduced in dry years compared with wetter years (through effects on survival, leaf area, etc.) much more than ε_s (Fig. 3.11a). In one year, however, much more rain fell during expansion than after. The canopies achieved a large leaf area before the full effects of

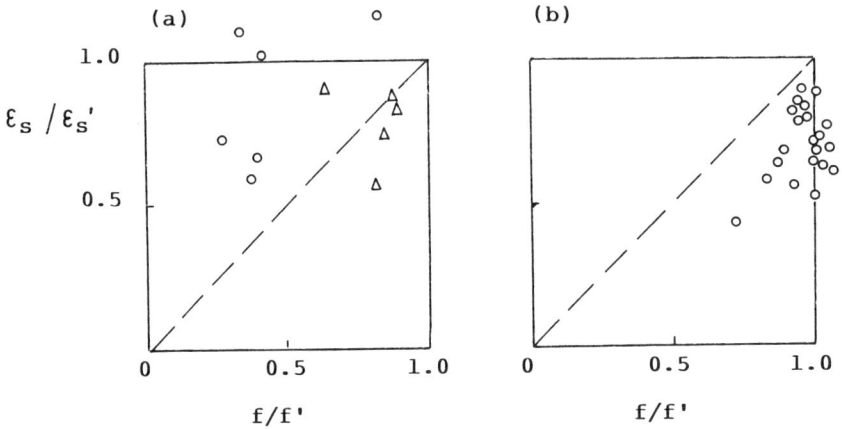

Fig. 3.11. **Contrasting responses to environmental stresses (mainly drought) for (a) sorghum in Botswana and (b) oil palm in Malaysia.** Fractional interception (f) and the conversion ratio (ε_s) for several stands are expressed in relation to the values of the stand that produced most dry matter (f', ε_s'). The sorghum stands were grown over four years in very dry conditions; usually f was reduced relatively more than ε_s, except in a year ($\triangle$) when early rainfall allowed rapid canopy expansion (source: Rees, 1986c). The oil palm stands were grown at sites differing mainly in rainfall and soil depth: invariably ε_s was reduced relatively more than f (source: Reference 25).

the drought were felt, and at harvest, ε_s for most of the stands had been reduced more than f. Generally, ε_s is more likely to be conserved if drought is experienced during, or relieved after, canopy formation.

In contrast to these annuals, the perennial oil palm responds to a wide range of dry and otherwise adverse environments by reducing ε_s much more than f. The stands in Nigeria referred to earlier (Fig. 3.6) suffered a prolonged dry season; those in Malaysia did not. Leaf area index (not shown) differed little between the sites, but ε_s was much smaller in Nigeria. In a more extensive set of trials in Malaysia[25], f hardly changed while ε_s was reduced in dry parts of the country to 50% of the maximum (Fig. 3.11b).

The influence of population (N_p) and intercropping

In semi-arid regions, stands are habitually established at low populations or as mixtures of species, and there has been some systematic study into whether these practices improve yield (by affecting seasonal interception or the mean conversion ratio). The evidence, though limited, indicates the effects on total production are small.

A range of responses to N_p was displayed in different years by a single cultivar of sorghum grown in Botswana between 1 and 21 m^{-2}[22]. When there was about 600 mm of rain in the season, dry matter production increased with N_p over the whole range. With about 200 mm, production increased with N_p up to about 10 m^{-2} then remained steady. With only 100 mm, production was similar over the whole range of N_p. As described in the previous section, these stands responded to low rainfall by variously reducing f and ε_s, but there was no indication in any of the dry years that more dry matter was produced at low than at high population. However, in an experiment with pearl millet in Niger, there was an apparent optimum population above which production decreased substantially[23]. Changes in both f and ε_s (above-ground material) caused the fall in production above the optimum, but the effect on ε_s might have been spurious. There was evidence that a larger fraction of the dry matter was allocated to roots deep in the soil, whose mass was not measured.

It is also uncertain that intercrops in very dry environments can intercept more solar energy or convert it more efficiently than their components grown alone. The taller crop is usually much sparser than it would be in moist conditions and transmits more solar radiation to the shorter. There might not be such a large increase in the conversion ratio of the shorter C3 as in moist conditions. In the work in Botswana[22], intercrops of sorghum and cowpea were also grown in two very dry years: the intercrop consisted of additional rows of cowpea sown between rows of sorghum (wide and narrow row-spacings). In neither year was the effect of intercropping statistically significant.

Dry matter production and plant water status

In this and the previous chapter, it has been indicated several times that processes that determine dry matter production are usually weakly related to water potentials in the plant. For many stands, including the millet and groundnut described in this chapter, leaf water potential is generally lower in dry than in moist environments, but a fall in potential does not have a consistent effect on photosynthesis and leaf area. For other stands, dry matter production seems unrelated to leaf water potential[27]. Only rarely do photosynthesis and leafing attributes seem consistently related to leaf water potential during much of the life-cycle: dry matter production, like leaf area, can then be expressed as an integral of time and water potential, as for cowpea in California[26].

These inconsistencies between stands could have several causes. The rates of most processes that govern f and ε_s probably depend more on the turgor potential in some part of the plant than on the corresponding water potential. Turgor potential commonly falls with a decrease in water potential, but can be maintained if the solute potential increases (for example, during both a day and an extended drought). The extent of this change in the solute potential – 'osmotic adjustment'[28] – differs much between genotypes and even between stands of a genotype, and can be large – up to 0.6 MPa in rice, for example.

Moreover, many physiological attributes seem not to depend on the 'bulk' potentials measured on the whole leaf or a large part of it. The leaf conductance and the rate of photosynthesis are influenced by the turgor of the cells immediately surrounding the stomata, which might be very different from the bulk potential. The rate of leaf extension is sometimes more closely coupled to the bulk potential (presumably because it is similar to that of the meristem). But processes such as leaf extension and photosynthesis also seem influenced by the degree of dryness experienced by other parts of the plant, particularly the root system[29]. In some circumstances, the bulk potential differs little between droughted and control stands producing dry matter at very different rates[27].

The prevalence and importance of these different forms of control is uncertain. At present, it is difficult to interpret and model the effects of drought on the many processes that determine the interception and conversion of solar radiation by stands in the field.

Effects of nutrient shortage

Little is known of the systematic effects of nutrients and other soil factors on fractional interception and the conversion ratio. Nutrients added to infertile soils usually increase plant production: the response is commonly asymptotic, but production sometimes decreases if very large amounts of nutrients are added[30]. (The response to nitrogen is sometimes small for legumes with root nodules that fix atmospheric nitrogen.)

Relations with leaf nutrient content

There have been few attempts to relate f and ε_s to the amount of nutrients in leaves. The nutrients can be expressed as a percentage concentration or a specific nutrient content (i.e. mass of nutrients per unit leaf area). Fractional interception seems weakly related to nutrient content, largely for the reasons considered in Chapter 2 (leaf expansion). In contrast, the photosynthesis rate and ε_s are strongly related to nutrient content[31]. In Figure 3.12, for example, ε_s for oil palm at a site in southern peninsular Malaysia increased in relation to potassium content of the fronds. Similarly, in northern Australia, ε_s for maize and sorghum responded positively and linearly to the specific leaf nitrogen (from e = 0.5 g MJ^{-1} at 0.6 g (N) m^{-2}, to e = 2.0 g MJ^{-1} at 1.7 g (N) m^{-2}). This range was achieved by applying different amounts of nitrogen fertilizer to the soil. The ratio ε_s differed between treatments both during expansion of the canopy, and after flowering when ε_s declined as N was translocated out of the leaves. When no nutrients were added, the amount of nitrogen that was retranslocated was so great that ε_s almost reached zero[32].

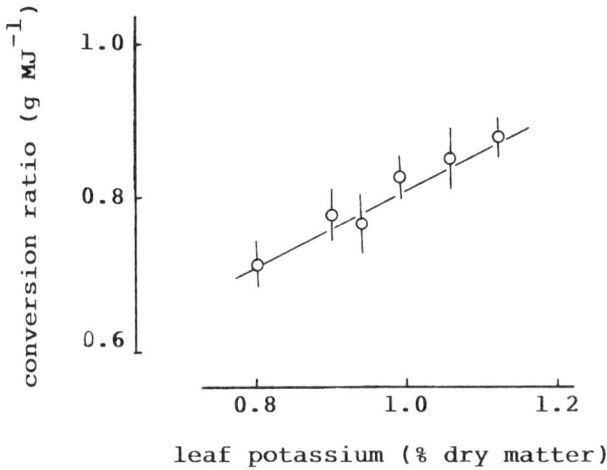

Fig. 3.12. Relation between the conversion ratio (ε_s) for photosynthetically active radiation and leaf potassium content in oil palm. From a large fertilizer trial in Malaysia. Values of ε_s for plots averaged in groups of 7 in ascending order of K content ($\pm$standard error)[25].

Such responses of f or ε_s to nutrient content are seldon consistent between sites, even for one genotype. For example, many other stands of oil palm, having a leaf potassium content within the range in Figure 3.12, have much larger values of ε_s than those shown. These stands are generally at moister sites, suggesting drought supressed ε_s for the stands in Figure 3.12 despite the high level of leaf potassium.

Relative effects on interception and conversion

In this work with maize, sorghum and oil palm, the nutrient status of the soil affected both f and ε_s. The responses of the two attributes are complicated and interdependent, but ε_s was suppressed more than f by scarcity of nutrients. Accordingly, when the dose of fertilizer was increased, f reached a 'plateau' at a lower dose than ε_s. This was more so in the perennial oil palm, which produces a large and stable canopy even when growing on infertile soils. Adding nutrients increased leaf area index somewhat, but had little effect on f. This was demonstrated by comparing values of L, f and ε_s in fully fertilized plots and control plots (no fertilizer) in 12 trials in peninsular Malaysia[33]. The mean responses to fertilizer (fully-fertilized/control) were 1.17 for L, 1.05 for f and 1.29 for ε_s[33].

The suppression of ε_s more than f by infertile soils, and conversely the greater stimulation of ε_s by fertilizer, was also demonstrated for maize and sorghum in the fertilizer trial in northern Australia[32]. The main effect of

(a)

(b)

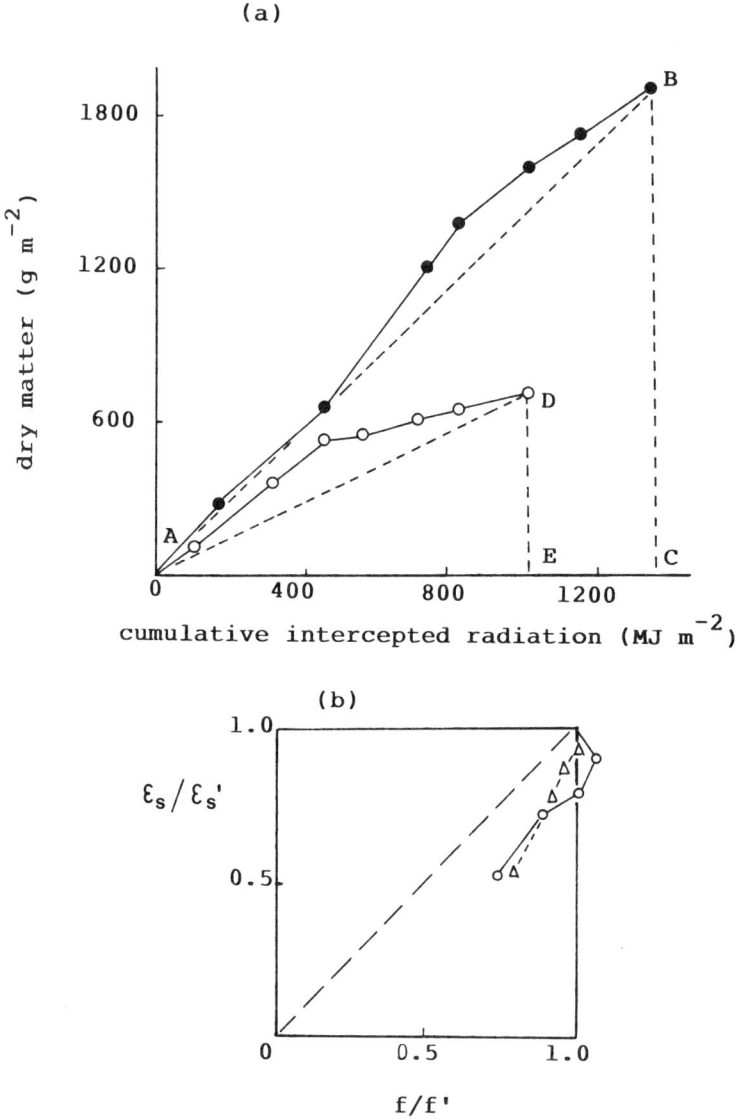

Fig. 3.13. Effects of nutrients on fractional interception (f) and the conversion ratio (ε_s) in maize and sorghum. From an experiment in Australia in which both species were grown at 5 levels of nitrogen fertilizer: (a) shows examples of dry mass in relation to cumulative intercepted (total) radiation for stands of maize (that produced the most (●) and least (○) dry matter, given $42 \, g \, m^{-2} \, N$ and zero N, respectively); in (b) stands are compared to the heaviest of each species (maize, ○ sorghum, △) in terms of f/f' determined as AE/AC, and $\varepsilon_s/\varepsilon_s'$ determined as (DE/AE)/(BC/AC). Decreasing amounts of nitrogen reduced ε_s more than f. From data in Muchow and Davis, 1988.

added nitrogen was to reduce the fall in specific leaf nitrogen during grain filling, and thereby to maintain ε_s at a large value throughout the stand's life (Fig. 3.13). The trial also suggested an important effect of population density. The sorghum was established at a higher population than the maize, and produced a larger leaf area in all treatments. The canopies of both species in most treatments were large enough to intercept most of the incoming solar radiation, so the extra leaf of the sorghum only slightly increased f. In each fertilizer treatment, the two species took up similar amounts of nitrogen from the soil, so the specific leaf nitrogen was greater for the maize than for the sorghum during the phase of expansion. As related above, ε_s was governed by the specific leaf nitrogen, and was therefore larger in the maize than the sorghum. The greater ε_s of the maize outweighed its smaller f, so the maize produced most vegetative dry matter. There were further effects during grain filling, when ε_s declined as nitrogen was translocated from leaves to grains. However, the maize had still produced more dry matter than the sorghum when the stands were harvested.

Concluding remarks

Though much is known of the way water and nutrients affect cellular processes, the knowledge has yet to be applied systematically to show how leaf area, fractional interception and the conversion ratio are variously controlled in different species. One of the main problems is relating an effect or response to an index within the plant, when the physiological systems are reacting to conserve the water status and nutrient content of the plant tissue. An alternative method of analysing dry matter production in dry conditions is examined in the next chapter, and the problems of a comparable analysis in nutrient-poor conditions are discussed in Chapter 6.

The stands in the various experiments referred to in this and the previous chapter responded by reducing both fractional interception (f) and the conversion ratio (ε_s), though some species reduced f most, others ε_s. The form of response usually taken by a species seems related to its general growth habit. The little information available suggests there is a range of responses, from that of the 'early' determinate herbs which reduce f and ε_s to about the same extent (the relative responses being governed by the form and timing of the environmental stress), to the indeterminate perennials that habitually reduce ε_s much more than f. The implications of the form of response for partition of dry matter between vegetative and reproductive structures are considered in Chapter 5.

References

1   Chapter 5 in Milthorpe and Moorby (General reading) gives an introduction to gas exchange and C3 and C4 photosynthesis. For the analysis using f and ε_s;

Warren Wilson, 1967, Monteith, 1972. Values of maximum production; Monteith, 1978 (and Further reading).

2 Field; Littleton *et al.*, 1981 (cowpea). Laboratory; El-Sharkawy and Cock, 1984, Veltkamp, 1985 (both cassava). For tree crops; Eckardt, 1987.

3 Kasanga and Monsi, 1954. For application; Littleton *et al.*, 1979b (cowpea), Hughes, Keatinge and Scott, 1981 (pigeon pea), Marshall and Willey, 1983 (millet/groundnut/intercrop), Muchow, 1985 (legumes).

4 Marshall and Willey, 1983. The factor of 1.4 was confirmed by the author in a comparison of fractional transmission measured by tube solarimeters and quantum sensors beneath oil palm in Malaysia. Equation 3.3 can be used when converting a value of ε_s based on total radiation to one based on PAR; the fraction of PAR in total irradiance is commonly assumed to be 0.5.

5 Hayashi and Ito, 1962.

6 Bell, Muchow and Wilson, 1987 (groundnut, Australia), Hughes, Keatinge and Scott, 1980 (pigeon pea, Trinidad).

7 Calculated from data in: Natarajan and Willey, 1980b (sorghum and pigeon pea), Reddy and Willey, 1981 (pearl millet and groundnut).

8 Unpublished data of Dr B. Marshall (summarized in *Microclimatology in Tropical Agriculture*, referred to in Introductory remarks).

9 For energy equivalents of dry matter, Westlake, 1963. For different approaches with an oil crop; Breure, 1988a, Corley, 1986.

10 Respiratory requirements; see articles by F. W. T. Penning de Vries (both 1975): The cost of maintenance processes in plant cells, in *Annals of Botany* **39**, 77–92 and Use of assimilates in higher plants, in J. P. Cooper (Further reading, this chapter). For application; see Breure, 1988b, who gives references to earlier work (e.g. by K. J. McRee, see also his 1974 paper in the main list).

11 Littleton *et al.*, 1981 (cowpea, Nigeria), Rawson and Constable, 1980 (pigeon pea).

12 For the main developments in the oil palm story; Rees, 1963, Corley, 1973, Breure, 1988a, b. The analysis summarized in Figure 3.6, and the data on oil palm in Figure 3.2 and Tables 3.1 and 3.2, are from the report, Squire, G. R. 1984, *Light interception, productivity and yield of oil palm*. Palm Oil Research Institute of Malaysia, Kuala Lumpur, p. 72.

13 Population and the conversion ratio; Bell, Muchow and Wilson, 1987 (groundnut), Hughes and Keatinge, 1983 (pigeon pea), Squire, 1984 (see Reference 12). See also Figure 2b in the paper by Williams, Loomis and Lepley, 1965 (corn in California, USA).

14 Ong and Monteith, 1985.

15 Matthews *et al.*, 1988b (groundnut), 1990b (sorghum), Veltkamp, 1985.

16 Intercropping; Natarajan and Willey, 1980b, Sivakumar and Virmani, 1980, Reddy and Willey, 1981 (all at ICRISAT, India). For analysis of light interception; Marshall and Willey, 1983. For shade studies on groundnut, Reference 14, Chapter 2; and cassava, Fukai *et al.*, 1984 (ε_s not given, but can be estimated from solar radiation, L, extinction coefficient, etc).

17 Littleton *et al.*, 1981. For gas exchange and temperature (growth chambers), see McRee, 1974.

18 Berry and Bjorkman, 1980 (review of thermal adaptation of photosynthesis).

19 Chamberlin and Songchao Insomphun, 1982.

20 For general discussion of nutrient uptake from the soil solution, see Chapters 3 and 4 in *Russell's Soil Conditions and Plant Growth*, 11th Edition (reference in Introductory remarks).

21 Matthews *et al.*, 1988b (groundnut), 1990b (sorghum), Shackell and Hall, 1979 (cowpea), Turner *et al.*, 1986a (rice).
22 Rees, 1986a, c.
23 Azam-Ali *et al.*, 1984a, b (pearl millet, showing apparent optimum population), 1989 (groundnut, showing effect on apparent ε_s of more dry matter in roots).
24 For example El-Sharkawy and Cock, 1984, El-Sharkawy, Cock and Held, 1984, Rawson and Constable, 1980.
25 Original measurements by various plantation agencies in Malaysia, collated by the agronomy unit at the Palm Oil Research Institute of Malaysia (PORIM), and analysed by the author. The data summarized in Figure 3.12 were collected by the Dunlop Research Department, Malaysia, and presented in this form by the Agronomy Unit, PORIM.
26 Turk and Hall, 1980b.
27 For an example from the field; Connor and Palta, 1981.
28 Turner *et al.*, 1986b (see also Further reading).
29 For a general discussion of this subject, see review by N. C. Turner, Further reading this chapter.
30 For a discussion of, and references to, the relation between applied nutrients and dry matter production, see pages 58–65 in *Russell's Soil Conditions and Plant Growth*, 11th Edition.
31 Lugg and Sinclair, 1981 (soybean), Murata, 1969 (rice), see also Sinclair and de Wit, 1976.
32 Muchow and Davis, 1988 (maize and sorghum)
33 Squire, 1986.

Further reading

Two chapters in the book edited by Cooper compare productivity by different stands in different environments. Monteith's 1978 paper (main reference list) is an amusing and instructive analysis of claims of maximum productivity. Corley's 1986 paper (main reference list) applies the analysis in this chapter to data for several plantation crops. On solar radiation and vegetation, Monteith's 1972 paper, is one of the standards, but less mathematical is the chapter by Monteith and Elston referred to in Further reading for Chapter 2. The symposium volume *Plant Canopies: their Growth, Formation and Function*, also referred to in Further reading, Chapter 2, has several relevant articles. The many effects of drought on physiological processes are discussed by Begg and Turner and in the sequel by Turner. Osmotic adjustment is reviewed by Morgan.

Begg, J. E. and Turner, N. C. (1976) Crop water deficits. *Advances in Agronomy*, **28**, 161–217.
Cooper, J. P. (1975) Control of photosynthetic productivity in terrestrial systems. In: Cooper, J. P.(ed.) *Photosynthesis and Productivity in Different Environments*. Cambridge University Press, London, pp. 593–621.

Loomis, R. S. and Gerakis, P. A. (1975) Productivity of agricultural ecosystems. In: Cooper, J. P. (ed.) *Photosynthesis and Productivity in Different Environments*. Cambridge University Press, London, pp. 145–72.

Monteith, J. L. 1972. Solar radiation and productivity in tropical ecosystems. *Journal of Applied Ecology*, **9**, 747–66.

Morgan, J. M. (1984) Osmoregulation and water stress in higher plants. *Annual Review of Plant Physiology*, **35**, 299–319.

Turner, N. C. (1986) Crop water deficits: a decade of progress. *Advances in Agronomy*, **39**, 1–51.

Chapter Four

Transpiration and Dry Matter Production

Earlier chapters considered how the environment influences the formation of a leaf canopy and a root system, and how the canopy intercepts and stores solar energy in dry matter. When water is limiting, stands intercept less radiation, and produce less dry matter per unit of radiation intercepted. The physiological mechanisms by which dry environments reduce dry matter production are not so well understood that dry matter production can be adequately modelled when drought is severe. Progress is now being made by taking an alternative approach, based on the amount of water a stand loses through transpiration.

General considerations

In much of the tropics, the amount and distribution of rainfall set the limit to plant production. The fraction of the rainfall that penetrates the soil and remains within the reach of root systems is so variable that production is much more closely related to evaporation than to the rainfall itself. For a group of stands in an experiment, the relation between dry matter and cumulative evaporation is usually linear, and can be defined by a slope and an (extrapolated) intercept on the evaporation axis. The values of slope and intercept are influenced by many factors, including the atmospheric humidity, the type of plant (e.g. Fig. 4.1a) and cultural conditions such as plant population density and the nutrient status of the soil (Figs 4.1b and c). Conditions that increase transpiration and reduce evaporation of water from other sources in the stand tend to increase the slope or decrease the intercept (or both). If these other losses are measured, and dry matter related to transpiration only, the intercept is negligible, but the slope is still affected by many physical and physiological factors.

The principles linking plant production with evaporation will now be examined. This first section considers the components of evaporation and the relation between dry matter and transpired water. A second section examines the factors controlling the rate and duration of transpiration, and hence the amount of transpired water. A final section summarizes the limitations to production in the dry tropics caused variously by the factors of soil, air and plant, and indicates some applications of the analysis. With one exception, the examples in the figures and tables are from experiments in the field.

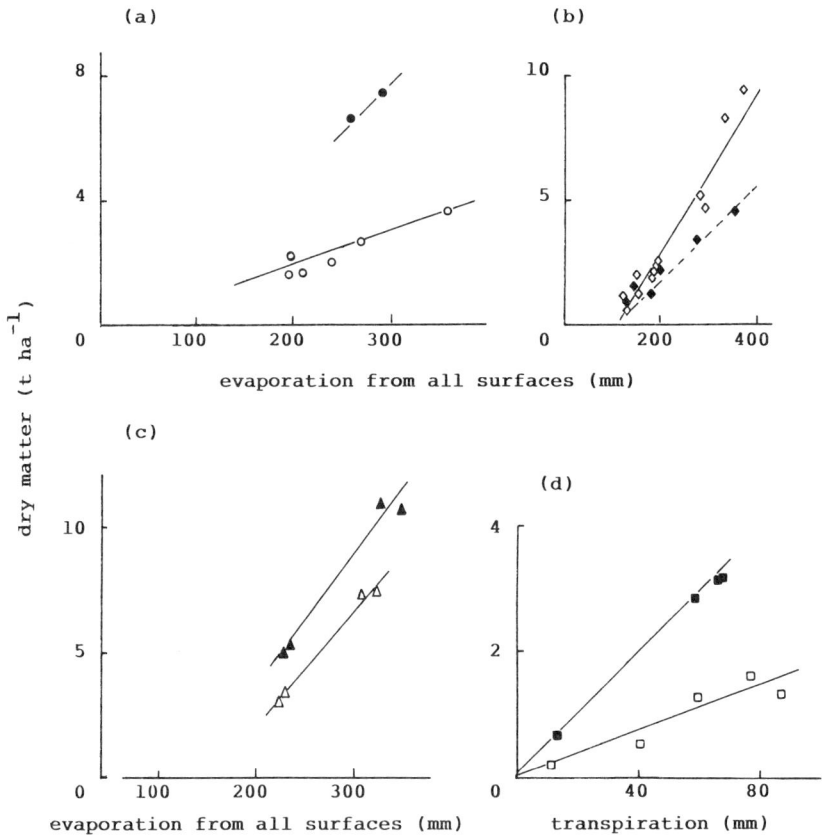

Fig. 4.1. Examples of dry matter production and evaporation in dryland agriculture. Parts (a) to (c) show values at harvest, and evaporation from all surfaces; part (d) shows values during growth in relation to transpiration only. (a) two C4 (●) and 6 C3 (○) species in the Philippines, from Angus *et al.*, 1983. (b) sorghum in Botswana (several years) at populations of 1.3 m^{-2} (◆), and 5.3 m^{-2} and greater (◊), from Rees, 1986c. (c) barley at two sites in Syria, with (▲) and without (△) fertilizer, from Cooper, Gregory, Keatinge and Brown, 1987. (d) pearl millet at Hyderabad, India (■) from Squire *et al.*, 1984, and in drier conditions at Niamey, Niger (□), from Azam-Ali, Gregory and Monteith, 1984b. Text gives explanations.

The components of evaporation

Measured rates of evaporation are commonly presented as a fraction of the potential evaporation rate (E_o), which is determined mainly by the weather, and estimated from a formula, such as Penman's for an open water surface[1]. Here, the main components of evaporation are identified by the following symbols:

E_s evaporation directly from the soil surface,
E_i evaporation of water intercepted by the canopy,
E_t transpiration through plants, per unit field area;
E_a total actual evaporation from all sources (crop, soil, standing water, etc.).

Techniques

Several techniques are used to estimate E_a and its components[2]. At the scale of the catchment, and sometimes the field, E_a can be estimated as a residual in the hydrological balance, i.e. precipitation minus other losses[2]. At the scale of the field (and over a period of at least several days), E_a is most commonly determined from the change in the water content of the soil, measured with a neutron moisture meter and tensiometers[4–8]. This technique allows drainage of water to be separated from upward fluxes (E_t and E_s) but the rainfall must be measured, and other losses such as surface runoff measured or eliminated. Over much shorter periods of time, E_a can be derived from the profiles within and above the stand of water vapour and other micrometeorological variables[3]; or with lysimeters, which measure changes in the weight of a mass of soil and plants[9].

Evaporation from the soil surface or from standing water (E_s) can be measured using small containers[6,10], such as metal cooking pots filled with soil or water and placed with their rim at the level of the evaporating surface. These containers are necessarily quite shallow, because they have to be removed to be weighed daily, and so can measure evaporation from soil accurately only during the first few days after a soil is wetted. Over longer periods, the method underestimates E_s by the amount of water that would move to the surface of the soil, perhaps from as deep as 0.5 m in some soils. Over long dry periods, E_s can be determined from the change in the water content of the surface layers of soil (measured by a neutron moisture meter, or by gravimetric sampling[5,6]). When a canopy is present, E_s must first be measured during drying of a bare part of a field. The amount of water lost from the soil beneath the canopy is then approximately the respective value of E_s for bare soil multiplied by the fraction of the incoming solar radiation reaching the soil through the crop [5]. Invariably, E_s decreases with time, such that cumulative evaporation is approximately a linear function of the square root of time since the last wetting – a relation that, once established for a soil, can be used to model E_s.

Interception of rain-water by the canopy is most commonly estimated as the difference between rainfall and the water falling through the canopy and flowing down stems (for example Reference 2). All these factors are spatially variable, and the estimate is usually uncertain. Losses from interception can also be estimated by the micrometeorological method, as the difference in evaporation rate of a wet and dry canopy[2,3]

Transpiration is commonly determined from soil water measurements or lysimetry as a difference, i.e. $E_a - E_s - E_i$, but can also be measured directly, at least when leaf surfaces are dry, over a much shorter period of time (minutes to hours), using a diffusive resistance porometer and ancillary equipment[12]. The transpiration per unit leaf area (E_l), is given by

$$E_l = (v_i - v_a)/(r_l + r_a) \tag{4.1}$$

where v_i and v_a are, respectively, the concentrations of water vapour in the intercellular spaces of the leaf and in the surrounding air, r_l the leaf diffusive resistance and r_a the leaf aerodynamic resistance. The vapour concentration v_i can be obtained from leaf temperature (using standard tables or a formula), on the assumption that the air inside a leaf is saturated with water vapour. (The leaf temperature is usually measured with a thermocouple or infra-red radiometer.) The atmospheric vapour concentration is most easily measured with an aspirated psychrometer held above the canopy. The resistance r_l is obtained with the porometer, and r_a from the loss of water from a leaf replica of moistened blotting paper. Transpiration rates are measured in several layers (or types) of foliage, weighted for leaf area, and summed to give E_t. This is the only method that can measure transpiration from flooded stands[10], or from the different components of mixed crops[12].

Typical rates of evaporation

The rate of evaporation from a stand depends on both the weather and the nature of the vegetation and soil. Examples of seasonal changes and totals of evaporation from different stands are given in Figure 4.2 and Table 4.1. (In Figure 4.2, total evaporation, E_a is expressed as a fraction of the potential, E_o.) In the tropics, perhaps only for inundated rice[10] and perennials in humid regions[2] is E_a/E_o near unity for much of the crop's life. Absolute values of E_a for such stands can be as large as 10 mm^{-1} (for example, for rice grown under clear skies as in Figure 4.3) but are usually smaller (4–6 mm d^{-1}) in the cloudy climates of the humid tropics. Most other stands lose water at a rate generally smaller than E_o.

For annual stands in moist conditions (Fig. 4.2a), E_a is typically at least 0.5 E_o when there is little foliage during expansion, since evaporation from

Table 4.1. Examples of evaporation from stands.

Species	Site	Duration	Cumulative values (mm)					Reference
			E_o	E_a	E_s	E_t	E_i	
Perennial plantation:								
Tea (dense canopy)	Kenya	365	1400	1250	s.	1050	200	Batchelor and Roberts, 1983
Annual, humid:								
Rice (paddy)	Sri Lanka	90	580	650	190	460	n.	Kowal and Kassam, 1973
Maize	Nigeria	113	570	480				Kassam, Kowal and Harkness, 1975
Groundnut	Nigeria	127	600	440				Kassam and Kowal, 1975
Pearl millet	Nigeria	84	400	345				
Maize/pigeon pea	India	98	540*	320	100	220		Sardar Singh and Russell, 1980
Annual, arid:								
Pigeon pea	India	119	560	250	30	220	n.	Sardar Sing and Russell, 1980
Cotton	Yemen	145	850	330			n.	Williams, 1979
Sorghum	Botswana	120	700	190	98	94	n.	Rees, 1986c

*Open pan evaporation.
Values are much influenced by site (soil, rainfall, etc.), and should not be considered representative of species. Symbols: E_o, potential evaporation, estimated by one or other of the modifications of Penman's formula; E_a, actual evaporation from all sources; E_s, evaporation from the soil surface or standing water; E_t, transpiration from plants; E_i, evaporation from water on plant surfaces; s., small; n., negligible. The dry season pigeon pea was part of the wet season maize/pigeon pea intercrop.

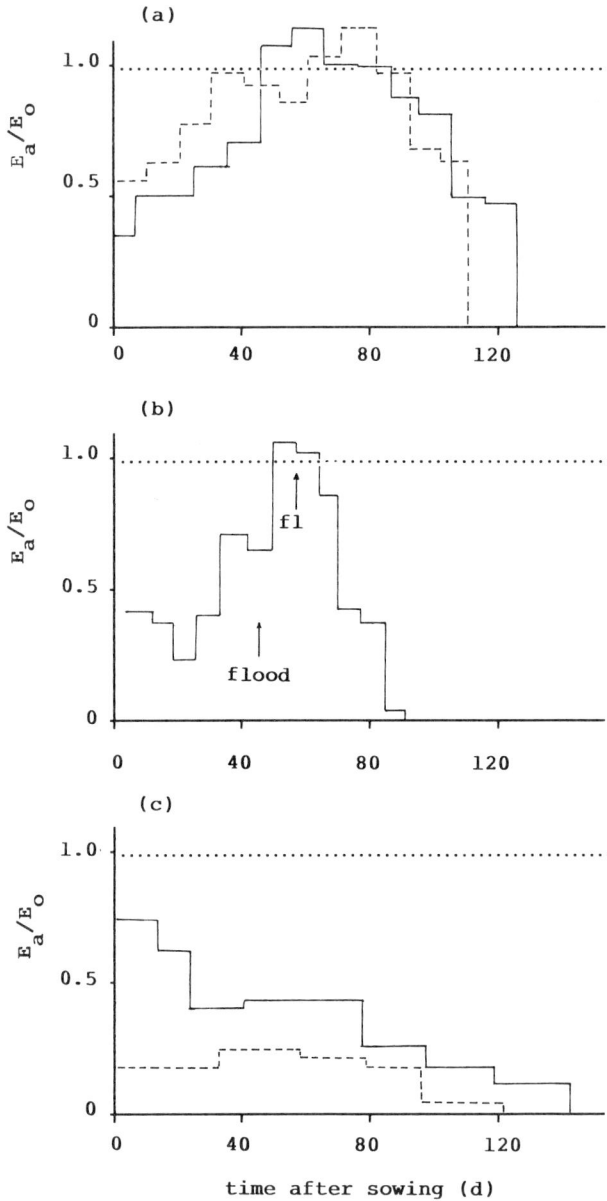

Fig. 4.2. Examples of seasonal change in evaporation from stands. Evaporation (E_a) from all surfaces expressed as a fraction of potential evaporation (E_o) derived from Penman's formula. E_a measured by weighing lysimeter in (a), estimated from changes in soil water content (neutron probe) in (b) and (c). (a) groundnut (——) and maize (– – – –) grown in different years in wet season at Samaru, Nigeria, from Kassam, Kowal and Harkness, 1975, and Kowal and Kassam, 1973. (b) pearl millet in the Yemen, irrigated at sowing, little rain, but receiving a flood irrigation at 45 d; flowering indicated by arrow marked f1; from Williams, 1979. (c) cotton in the Yemen (——), irrigated at sowing, little additional rainfall, from Williams, 1979; and sorghum in Botswana (– – – –), sporadic rainfall only, from Rees 1986c.

a moist soil will compensate for the incomplete canopy. (Total losses by E_s can be 0.3–0.5 of cumulative E_o until the canopy expands enough to cast a heavy shade on the soil.) The full canopy can itself lose water at a rate close to E_o (or even slightly faster if it is aerodynamically 'rough'), especially if frequently wetted; but within a short time of the canopy being fully formed, and usually when reproductive growth has begun, E_t usually decreases as a result of senescence of foliage, and by harvest has typically decreased to 0.5 E_o. At that time, the total loss will be little greater than this if many dead leaves remain on the plant and restrict evaporation from the wet soil. The mean rate for such annuals is usually less than for perennials, because of the slower rates during formation and senescence of the canopy, and is typically 0.4–0.6 E_o. The C4 and C3 species differ little in the maximum rate of evaporation.

Frequently irrigated stands can lose water at least as rapidly as those in moist environments. The total loss, E_a, from small irrigated plots in an extensive dry area can even be greater than E_o if additional energy for evaporation is supplied from the surroundings (advection)[13]. Where a stand grows partly or largely on little rainfall or the store of water in a soil, the combined losses seldom approach E_o (Fig. 4.2b, c).

The contributions of E_i and E_s to the cumulative total loss depend much on the frequency of rain. A complete canopy typically intercepts about 1 mm of rainfall, so in the example of tea in Table 4.1, with 200 rain-days in a year, the cumulative interception is 200 mm – 16% of cumulative E_a and 14% of E_o. In wet seasons, interception by most canopies will be between 0.1–0.2 E_o, but is seldom accounted for in estimates of E_t. Evaporation from a permanently wet soil or from the water of a rice paddy may be as much as 30% of the total loss during the life of the crop[10], though a smaller proportion – about 10% – is commonly lost from the moist soil beneath a perennial stand. The total losses by E_i and E_s during wet seasons therefore range from 0.2–0.5 E_o, and transpiration ranges accordingly from 0.5–0.8 E_o.

In dryland agriculture, where the canopy is much less frequently wetted, the loss by E_i is very small, but the loss by E_s can still be large. If rain falls frequently in small amounts, and especially if the stand is sparse, E_s can constitute 50% or more of the total loss[5,7]. However, if the stand grows mainly on water stored in the soil from previous rain or irrigation E_s will decrease rapidly as the soil surface dries, and might be no larger than 10–20% of the cumulative E_a[6]. In these dry environments, transpiration through the canopy might reach 0.4–0.6 E_o, or occasionally greater during the few days immediately after irrigation or rain (e.g. Fig. 4.2b) but the average value of E_t from emergence of the seedlings to the time transpiration ceases is usually little more than 0.1–0.2 E_o.

Among the stands in Table 4.1, the effects of crop duration added to the differences in transpiration rate which resulted in a more than tenfold

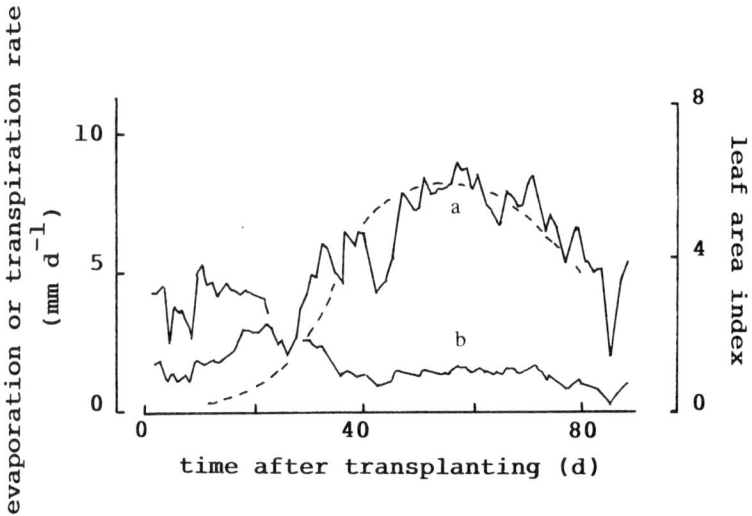

Fig. 4.3. Seasonal change of evaporation from a stand of inundated rice in Sri Lanka. Showing transpiration from the leaves and panicles, determined by porometry (a), evaporation from the irrigation water (b), and leaf area index (– – – –). From Batchelor and Roberts, 1983.

range of variation in cumulative E_t. The tea canopy transpired the most water – 1050 mm over a year when E_t/E_o was 0.75 and mean E_o was 3.8 mm d^{-1}; the dryland sorghum transpired the least – 90 mm over 120 d, when E_t/E_o was 0.13 and E_o was 5.8 mm d^{-1}. The rest of this chapter concentrates on transpiration and its effects at the dry end of this range, where crops grow on an uncertain and usually small amount of rain.

Dry matter and transpired water

In the analysis here, the dry weight, or dry mass, of a stand (W) is related to the cumulative total of transpired water by

$$W = \varepsilon_w \Sigma E_t \qquad (4.2)$$

where ε_w is the amount of dry matter produced per unit transpired water (the dry matter/transpired water ratio). As in Figure 4.1d, W generally increases more or less linearly with ΣE_t, such that ε_w changes little during growth[14]. The relation in equation 4.2 has its basis in the gas exchange of leaves, specifically that water is lost from leaves by transpiration whenever stomata open to allow carbon dioxide to diffuse from the atmosphere to the chloroplasts.

The basis of the dry matter/transpired water ratio: gas exchange

The net uptake of carbon dioxide (A_l) by a sunlit leaf can be represented by

$$A_l = (c_a - c_i)/r \qquad (4.3)$$

where c_a is the concentration of CO_2 in the air surrounding the leaf, c_i is the concentration in the intercellular spaces, and r is the sum of the diffusion resistances to CO_2 provided by the boundary layer and the leaf cuticle. The loss of water by transpiration (E_l) from the same leaf can be represented by

$$E_l = (v_i - v_a)/r' \qquad (4.4)$$

where v_i and v_a are the internal and external concentrations of water vapour and r' the corresponding diffusive resistance (i.e. $r_a + r_l$ in equation 4.1). Consequently, the rates of net photosynthesis and transpiration are related by the ratio

$$A_l/E_l = (c_a - c_i)/\{\beta(v_i - v_a)\} \qquad (4.5)$$

where β – effectively a constant – is the ratio of the diffusion resistances for CO_2 and water vapour[15].

The atmospheric concentrations of CO_2 and water vapour differ in that c_a changes relatively little with time of day and with season and locality, but v_a changes considerably with atmospheric temperature and humidity. The internal concentration of CO_2 (c_i) although not constant, is also much less variable than the internal concentration of water vapour, which is controlled by leaf temperature (on the assumption that air in the intercellular spaces is always saturated with water vapour). The difference ($v_i - v_a$) is therefore much more variable than the corresponding difference for CO_2; and under natural conditions, the ratio A_l/E_l is not constant. When leaf and air temperature are the same – and they are often similar when a stand is transpiring rapidly – the difference ($v_i - v_a$) is proportional to the saturation vapour pressure deficit, D (Reference 24, Chapter 2); A_l/E_l should then be inversely proportional to D, irrespective of all other factors such as the absolute value of diffusive resistances (for example, Reference 16).

Determination of the ratio at the scale of the stand

The relation between A_l/E_l and D should have a considerable bearing on the production of dry matter by whole canopies, since these are composed of leaves whose gas exchange is governed by equation 4.5. However, several factors change with the change in scale from leaf to canopy. Neither v_a nor c_a are uniform within the canopy, and c_i may also differ

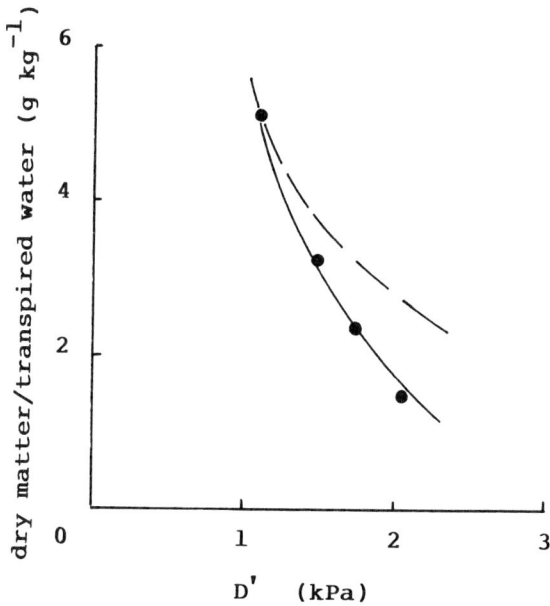

Fig. 4.4. The effect of saturation deficit, D', on the dry matter to transpired water ratio, ε_w. For stands of groundnut in controlled environment glasshouses, on a drying soil. The saturation deficit is the mean leaf-to-air value between 0500 and 1900 GMT. The dashed line is the relation based on the value of $\varepsilon_w D' = 5.6$ g kPa kg^{-1} in the most humid treatment. Source: Ong, Simmonds and Matthews, 1987.

between leaves and in response to environmental factors. More important, a large and variable proportion (about half for herbs) of the CO_2 assimilated is respired. Notwithstanding these complications, observations on many species have shown that the dry matter to transpired water ratio, ε_w, is inversely proportional to the saturation deficit, D (Fig. 4.4), such that $\varepsilon_w D$ is often conservative[15].

This effect was largely responsible for the differences between the stands in Figure 4.1d. For these and two others of the same genotype of pearl millet $\varepsilon_w D$ changed little, with a range of only $\pm 12\%$ of the mean (Table 4.2). Some of this variation was probably the result of systematic or random errors in measurement; others may have been due to different leaf and air temperatures, or because D was a daily maximum rather than a daily mean over the relevant hours. Nevertheless the conservatism of $\varepsilon_w D$ implies that v_a and c_a varied little within a stand. It also suggests that c_i was similar between stands and the proportion of assimilate respired was

Table 4.2. Comparison of dry matter excluding roots (W), transpired water (ΣE_t), the dry matter-water ratio ($\varepsilon_w = W/\Sigma E_t$), saturation deficit (D) and ε_wD for stands of pearl millet and groundnut.

Crop	Stand	W ($g\,m^{-2}$)	ΣE_t ($kg\,m^{-2}$)	ε_w ($g\,kg^{-1}$)	D (kPa)	ε_wD ($g\,kPa\,kg^{-1}$)	Location
Pearl millet							
	1	1440	220	6.4	1.4	9.0	Glasshouses at Nottingham (see text).
	2	600	150	3.9	2.4	9.5	Hyderabad, dry season, irrigated: Squire *et al.*, 1984.
	3	320	70	4.6	2.3	10.6	Hyderabad, dry season, drying soil: Squire *et al.*, 1984.
	4	170	80	2.1	4.0	8.4	Niamey, dry season, drying soil: Azam-Ali, Gregory and Monteith, 1984.
Groundnut							
	5	270	52	5.2	1.0	5.0	Treatments in glasshouses at Nottingham, in which D was the main variable, drying soil: Ong, Simmonds and Matthews, 1987.
	6	250	76	3.0	1.4	4.1	
	7	200	82	2.6	1.6	4.0	
	8	110	72	1.5	2.0	2.9	
	9	220	110	2.0	2.1	4.2	Hyderabad, drying soil: Azam-Ali *et al.*, 1989.
	10	420	220	1.9	about 2.5	> 4.0, < 5.0	Hyderabad, dry season with occasional irrigation: Matthews *et al.*, 1988a.

Note: the saturation deficit, D, is a daily maximum for stands 1–4, and a mean during the daylight hours for stands 5–9. The daily maximum for stand 10 is 3.1 kPa, giving ε_wD = 5.9 $kPa\,kg^{-1}$, but if based on the corresponding daylight mean, ε_wD would be within the range indicated.

similar in the different environments (a conclusion consistent with that for the radiation conversion coefficient).

However, $\varepsilon_w D$ is sometimes reduced by limitations such as drought or nutrient shortage. The example in Figure 4.4 illustrates an effect of drought for stands of groundnut grown in controlled environment glasshouses with their roots in field soil. The soil was fully wetted initially, then stands were grown without further water at different saturation deficits. Leaf temperature was measured so that the leaf-to-air vapour pressure deficit, D' (comparable to the vapour concentration difference $v_i - v_a$ in equation 4.5), could be determined for greater accuracy, and ε_w was based on the dry matter above ground. Although the value of ε_w decreased as D' increased, $\varepsilon_w D'$ was not conserved but itself decreased from 5.6 to 3.1 g kPa kg^{-1}. This might have been because more dry matter was allocated to roots in the drier treatments. The leaf water potentials however, were lower in the drier treatments and might have limited photosynthesis, thereby decreasing A_l/E_l by raising c_i in equation 4.5. Nutrient shortage probably affects ε_w also through c_i, but the general importance of this response in the field is uncertain (for example see p. 137).

Differences between C3 and C4 species

When grown in similar atmospheric environments, the tropical cereals with the C4 photosynthetic pathway commonly have a dry matter to transpired water ratio just more than twice that of C3 species – a difference of the order expected on the basis of the typical difference in c_i/c_a (equation 4.8) between C4 and C3 plants[15]. In the comparison of several species in the Philippines (Figure 4.1a[17]) ε_w was 2.2 times greater for maize and sorghum than for C3 stands. (Here, ε_w was based on a glucose equivalent to account for differences between the species in the energy content of their fruiting structures.) The value of $\varepsilon_w D$ differed similarly between the pearl millet and groundnut in Table 4.2. In moist conditions (stands 1, 2, 5, 6 and 7), $\varepsilon_w D$ was 2.1 times larger for millet than groundnut (based on dry matter). However, the overall means of $\varepsilon_w D$ (9.5 g kPa kg^{-1} for millet and 3.7 g kPa kg^{-1} for groundnut) differed by a factor of 2.6: the dry conditions seem to have reduced $\varepsilon_w D$ for groundnut (as in Fig. 4.4) but not for millet.

The influence of D over ε_w is one of the most important effects restricting dry matter production in dry regions, and is represented in Figure 4.5 for a typical amount of extracted water. The value of $\varepsilon_w D$ for the millet is the mean from Table 4.2; the two values for the groundnut are those in moist and dry conditions in Table 4.2, between which the leaf water potential differed by about 0.5 kPa. However, it is not certain that C4 species always have a larger dry matter to transpired water ratio than C3 species. The ratio is sometimes similar for drought tolerant C3 species

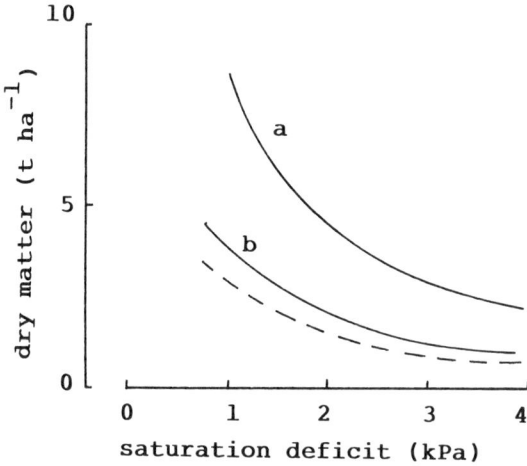

Fig. 4.5. Modelled responses of dry matter production (W) to saturation deficit (D). For (a) pearl millet and (b) groundnut. Curves show the dry matter from 100 mm of extracted water using a dry matter/transpired water ratio, ε_w. The dashed curve shows the response for groundnut when ε_w is further reduced by a decrease of about 0.5 MPa in mean daytime leaf water potential (text gives explanation).

such as cotton and cowpea, and relatively drought sensitive C4 such as maize and some sorghums (for example, Reference 5). It is possible that adverse conditions affect ε_w (through c_i) much more in maize, than for example cotton, and cancel the C4 advantage.

Factors controlling transpiration

It was shown earlier that attributes of the vegetation, rather than the atmosphere, normally limit the rate at which water is transpired from most dryland crops. These attributes of the vegetation can be grouped into two categories: those of the canopy determining the movement of water vapour from the sub-stomatal cavities of leaves to the turbulent air above the canopy; and those of the root system that determine the supply of water from the soil to the conducting vessels in the plant. These attributes cannot be independent: the rate of loss from the canopy must equal the rate of extraction by the root system. However, it is usually not clear whether the root system or the canopy primarily controls the rate of transpiration at any time, and which limits the total volume of water removed from the soil. The factors of the canopy and root system that control transpiration will now be considered.

Control of transpiration by the canopy

The structure and physiology of canopies influence transpiration through both the aerodynamic resistance for transfer of water vapour between the canopy and the atmosphere, and the combined diffusive resistances of the cuticle and stomata of the various leaves and other transpiring organs, r_c. This canopy resistance is increasingly used to estimate evaporation by a modification of Penman's formula[1,2,3].

In physiological analyses, the diffusive conductance ($g_c = 1/r_c$) is often preferred to resistance, since transpiration increases with increase of conductance over a wide range. The canopy conductance is usually estimated as the sum of the products of leaf area index (L) and leaf conductance (g_l) for a number of layers or types of foliage[18]. It can also be determined from the micrometeorological technique[3], provided the canopy is dry and soil evaporation negligible. If the former technique is used, the floral structures of certain species, specially large-panicled cereals, must be included since they might contribute as much as 30% to the canopy conductance and transpiration, and 15% to the total loss of water from the plants between sowing and harvest (e.g. rice[10]).

The aerodynamic conductance is proportional to windspeed, though affected somewhat by plant attributes such as the height and roughness of the canopy. It is important in controlling transpiration from deep, dense canopies, especially at low windspeeds. However, the canopy conductance is the main discriminant of transpiration from sparse, senescent or stressed canopies for which E_t is much smaller than E_o. In these circumstances, g_c can be an indicator of the approximate value of E_t/E_o. Stands of most crops need a canopy conductance of at least 1.5–2 cm s^{-1}, depending on the aerodynamic conductance, before they transpire near the potential rate. (In terms of canopy resistance, these values are 67–50 s m^{-1}.) A conductance of this order can be achieved, for example, when a canopy has a leaf area index of 3–4, and a leaf conductance averaged through the canopy of about 0.6 cm s^{-1}.

The canopy conductance of stands during senescence or in dry conditions is generally smaller than 1.5 cm s^{-1}. For example, g_c of irrigated millet was 0.5–1.0 cm s^{-1} during senescence when E_t/E_o was 0.5, and about 0.4 cm s^{-1} for a stand growing on a store of water when E_t/E_o was 0.3[19]. Such small canopy conductances are invariably the result of reductions in both leaf area index (Chapter 2) and mean leaf conductance.

Control of leaf conductance

Leaf conductance varies considerably within, and between, leaf surfaces and between parts of the canopy. It is influenced by developmental stage and responds to many environmental factors. (Reviews are given in Further reading, this chapter.) In moist conditions, g_l depends mainly on irradiance, except during senescence of the stand when it might be small, irrespective of the irradiance. The effects of dry conditions are complicated. Even when the soil is moist, g_l usually decreases as the saturation deficit increases[20] – a response that to an extent conserves the transpiration rate, and by doing so, reduces E_t/E_o. Figure 4.6 shows how such an effect of saturation deficit influenced the canopy conductance of tea, determined by a micrometeorological method. Here, g_c was also influenced by the irradiance, but the effect of D over the range shown was sufficient to reduce E_t/E_o by 20% (from about 0.75 to 0.55).

A drying soil suppresses g_l independently of any effects of the atmosphere, but during a long drought the effects of air and soil are very difficult to distinguish. Moreover, the responses to dry air and dry soil can

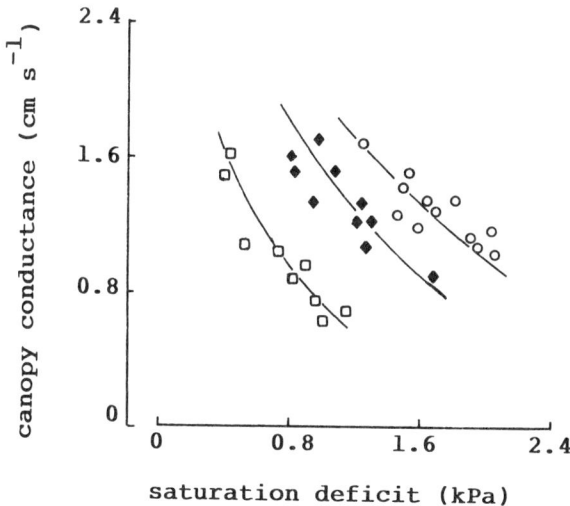

Fig. 4.6. The relation between canopy conductance and the saturation deficit of the air at the surface of a tea plantation. In the wet season in western Kenya, when the canopy was dry. Canopy conductance was measured by a micrometeorological technique (see text). Values of conductance are grouped in three intervals of net radiation: □, 100 to 150 W m^{-2}; ♦, 350 to 400 W m^{-2}; ○, 600 to 650 W m^{-2}. Adapted from Callander and Woodhead, 1981.

be modified by change in the ratio of leaf area to root length (as described later).

The physiological mechanisms mediating the responses to drought have been difficult to define. Leaf conductance is commonly little affected by the 'bulk' water potential of the leaf. In some circumstances, g_l has been found to be smaller in a droughted stand than a watered control, even though the leaf water potential was similar in both stands, or higher (less negative) in the droughted than the control[21]. Conductance is probably more sensitive to both the water and turgor potentials of the cells around the stomata, and these potentials might be very different from the bulk potentials. There is also evidence that g_l is influenced independently of conditions in the leaf by hormonal effects originating in the root system. Eventually, after a long period of drought, g_l and the bulk leaf water potential both decrease, but it is uncertain even then how they affect each other.

Limiting attributes: leaf area and leaf conductance

In moist conditions, E_t, or E_t/E_o, increases in proportion to the rise in leaf area index (L), as in Figure 4.3, and approaches unity when the canopy absorbs most of the incident radiation (typically when L is between 2.5 and 4, depending on canopy architecture, and the state of the foliage). The transpiration rate usually rises more rapidly for cereals than legumes (as in Fig. 4.2a), consistent with the difference in the rate of canopy expansion (Figs 2.6, 3.4). The effect of light on stomata allows E_t to rise with E_o in sunny weather, but the concomitant rise of saturation deficit (with rise in air temperature) sometimes restricts E_t below E_o. When the canopy senesces, E_t usually decreases through effects on both conductance and leaf area.

In Chapter 3, it was shown that most species respond to drought by reducing both leaf area and photosynthesis per unit leaf area. The effect on photosynthesis is usually accompanied (and caused) by an effect on conductance. As in the examples in Chapter 3, the extent to which area and conductance are reduced depends on both the timing of the drought and on the genotype of the crop (some tending to reduce conductance more than area, others area more than conductance). Groundnut – probably representative of many indeterminate herbs – restricts E_t more through leaf conductance than leaf area. Figure 4.7a, for example, shows the changes in a stand growing without rainfall on a store of water. For much of a period of 40 d, during which the canopy continued to expand, transpiration fell slightly: E_t was controlled by a threefold decline in mean leaf conductance. The responses of millet (not shown) differ from those of groundnut in that during expansion, leaf conductance is conserved more than leaf area.

These different forms of response can also occur between genotypes of species. For example, two cassava genotypes transpired at the same rate during a drought: they had a similar root length, but one – known to be vegetatively vigorous – had twice the leaf area of the other, but a much smaller conductance per unit leaf area[21].

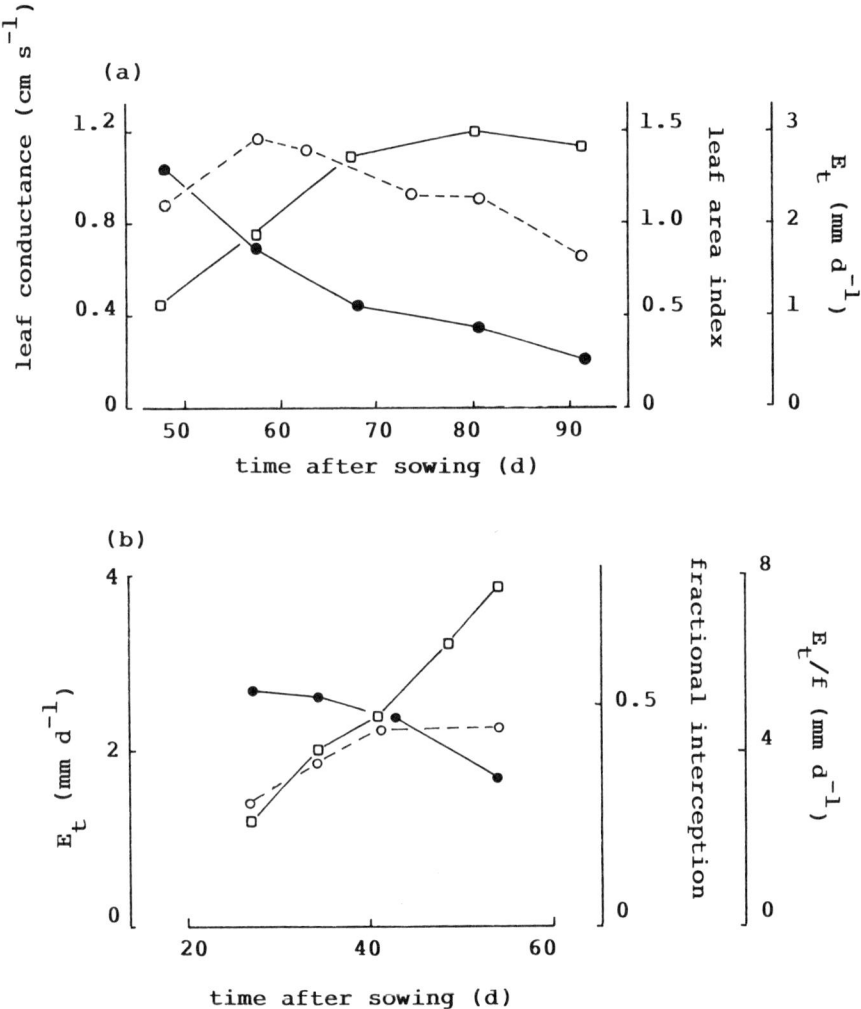

Fig. 4.7. Control of transpiration by leaf conductance. For stands of groundnut: (a) on a drying soil at Hyderabad, India, showing leaf area index (□) , mean leaf conductance (●) and transpiration rate per unit ground area E_t (○), adapted from Simmonds and Azam-Ali, 1989, with additional information; (b) in controlled environment glasshouses, showing the analysis in which transpiration rate, E_t (○), is normalized with respect to fractional interception of solar radiation, f (□) to give E_t/f (●), adapted from Simmonds and Ong, 1987.

Transpiration and intercepted radiation

It was clear in Figure 4.7a which attribute controlled transpiration, but the comparable analysis for other stands can be ambiguous. Luxuriant stands can lose much foliage while hardly affecting transpiration; some sparser canopies regulate transpiration by reducing the area of leaf effectively exposed to the sun by means of leaf movements. More generally, the contribution of leaf conductance to the control of E_t can be better revealed by expressing transpiration rate in terms of intercepted solar radiation – a procedure analogous to that for determining the conversion ratio (Chapter 3).

In moist conditions (irrigated soil, $D < 1.5$ kPa), most of the intercepted radiation is used for evaporating water: the amount of water transpired per unit intercepted radiation is a conservative quantity, generally similar to the reciprocal of the latent heat of vaporization of water (0.41 kg MJ^{-1} at 25°C). As stands senesce or become short of water, the water/radiation quotient decreases, suggesting a reduction in leaf conductance[6,22].

The analysis of this effect can be simplified for stands receiving the same incoming radiation, for example, if they form a series of treatments at one site. Transpiration rate can then be expressed in terms of fractional interception of solar radiation (f), rather than intercepted radiation itself[6]. In Figure 4.7b, the analysis of E_t/f is applied to a set of responses similar to those in Figure 4.7a. Fractional interception increased continuously, but the concomitant decrease in E_t/f (caused by a decrease in leaf conductance) was sufficient, by 40 d after sowing, to compensate for the rising f and to conserve E_t.

Limitations of the analysis for the canopy

The changes in area and conductance that regulate E_t are not always traced to effects of atmospheric environment or plant development. It is usually unclear why E_t in dry conditions is so much smaller than E_o, and why E_t sometimes falls to negligible values, after being stable for several weeks. A change of transpiration sometimes occurs when the atmospheric conditions are themselves stable and unlikely to bring about substantial change in leaf conductance, or when the canopy consists of young, green, expanding leaves. In these circumstances, the control of transpiration must be sought among the relevant factors of the root system and the soil.

Control of transpiration by the root system

The amount of water that a stand has access to depends on many factors, including the rainfall and the amount of water that can be stored in the soil. In dry regions, stands are sometimes planted at the end of the rainy season, or after a flood irrigation or short rains, and grow for most of their life on the store of water. Many other dryland crops grow for at least part of their life on a store of water. The amount that can be held in this store depends on the physical conditions of the soil, and the amount the stand can extract depends also on the physiological attributes of the genotype.

The depth of a soil and the amount of water it holds set the upper limit to the water available to a stand. Root systems are unable to extract all the water in a soil, so soils are generally compared in terms of the water they hold at tensions between a nominal 'field capacity' (e.g. 10 kPa), when the soil is mostly drained after a thorough wetting, and a 'permanent wilting point' (e.g. 1500 kPa). The volumetric water content between these limits – the available water – can be as large as 0.4 (40%) for some organic soils and as little as 0.07 for light sands. Among the loams and clay loams, it is typically 0.2–0.25. Those soils with more clay can hold more water, but plant roots can less easily obtain this water (since it is held at a higher suction); the more medium textured of the loams and clay loams therefore usually have the most available water[24].

Extraction of water by roots

Little is known of whether the different parts of the fine root system in the field extract water at different rates. In most analyses, the fine roots are unclassified, and the rate at which water is extracted from soil and supplied to the canopy is considered in terms of the total length of root (R, in equation 2.2) or the length per unit soil volume (l_v). A mean rate of inflow of water (I) per unit length of root is determined for the root system, or the roots in a layer of soil. Considering the whole system, for instance, the transpiration rate (E_t) can be expressed as

$$E_t = RI \qquad (4.6)$$

When the units of E_t are volume of water transpired per unit of ground surface per unit time, R is length of root per unit area of ground surface, and I is volume of water per unit length of root per unit time. The factors influencing the length of root were considered in Chapter 2. The inflow of the fine roots depends on several factors of the soil and plant, but particularly the gradient of water potential between the leaves and the soil adjacent to the roots[6,25].

In wet seasons, most stands produce enough root length to extract water at near the potential evaporation rate, which is typically 5 mm d^{-1}. For the monsoon stands of pearl millet and groundnut examined in Chapter 2, R was 2–3 km m^{-2}, and mean rates of inflow (averaged over about a week) must therefore have been 1.7–2.5 g m^{-1} d^{-1}. The much larger root systems of cassava and rice (see Fig. 2.13) still would have

time after sowing (d)

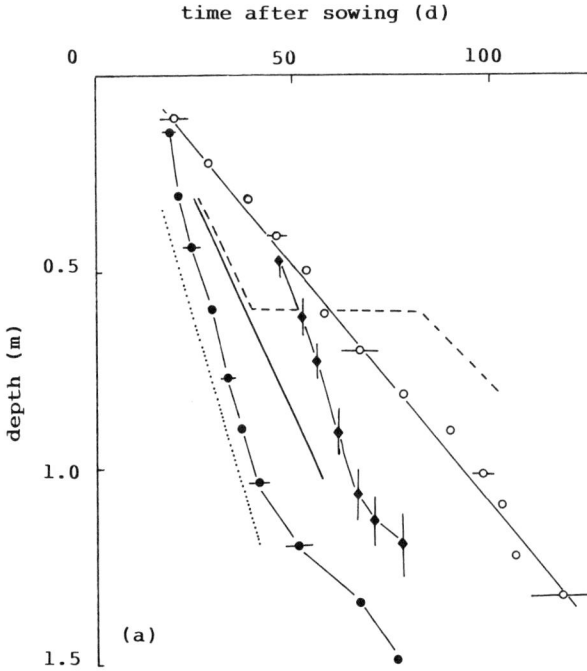

Fig. 4.8. Movement of the water extraction front below stands of several annuals. (a) at population densities between 10 and 29 m^{-2}, mostly in sandy loams and sandy clays; bars indicates representative ($\pm$) standard errors of either the mean time for the front to reach a given depth (horizontal bars), or the mean depth reached by the front at a given time (vertical bars):

....	millet at Hyderabad, India: Squire *et al.*, 1984.
•	sorghum at Hyderabad: Matthews *et al.*, 1990b
————	sorghum in Botswana, soil profile moist: Rees, 1986c.
– – – –	sorghum in Botswana, soil profile not fully moistened, extraction front ceased descending at 40 d, then continued after further rain at 80 d: Rees, 1986c.
○	groundnut at Hyderabad: Matthews *et al.*, 1988a.
◆	groundnut at Hyderabad: Simmonds and Williams, 1989.

transpired at a similar rate, so mean inflow would have been much smaller than the values just given. Inflow of all these stands, but particularly those of the cassava and rice would probably have been limited by the potential evaporation rate, rather than factors of the soil or root system.

In dry soils however, the mean inflow depends much less on the atmosphere. The transpiration (= extraction) rates for herbs such as millet

time after sowing (d)

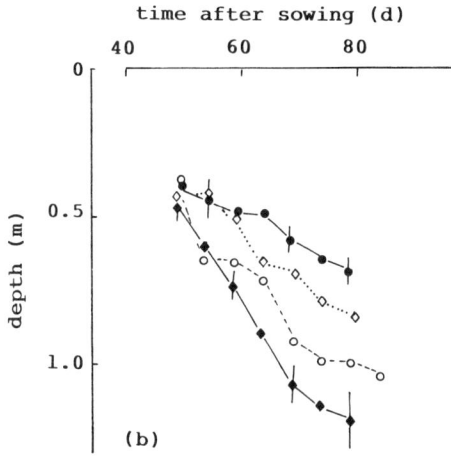

Fig. 4.8. contd. (b) for groundnut at Hyderabad, from Simmonds and Williams, 1989 (above) at populations of 22 (♦, the same line as ♦ in part a), 11 (○), 7 (◊) and 0.6 m^{-2} (●); within 0.2 m of the row, ±s.e.

and groundnut are typically 2–3 mm d^{-1} when growing on a store of water, whereas the total length of root is usually greater in these circumstances than when a soil is continuously moist throughout the profile. Therefore mean rates of inflow are smaller for root systems exploring a store of water. For example, a millet stand at Hyderabad extracting from the store had a total root length of 3.0 km m^{-2}, and transpired at 3.1 mm d^{-1}, equivalent to an inflow of 1.0 g m^{-1} d^{-1}. Corresponding figures for a stand in Niger were 5.9 km m^{-2} for root length, 3.8 mm d^{-1} for transpiration rate and 0.65 g m^{-1} d^{-1} for inflow In these examples, the mean inflow was one-half to one-quarter that in a moist soil – a result of the more negative soil water potential causing a decrease in the mean gradient of water potential between soil and plant.

Limitations to extraction from a store of water

Measurements of soil water content with a neutron moisture meter and soil water potential with tensiometers usually reveal an extraction front descending into a drying soil (for technique, see References 4, 6, 19, 23, in this chapter). Perennials with an established root system can extract from a wide depth-interval of soil above the front. (If rain wets the surface layers, the front stops moving until that water has been extracted.) Such perennial systems can extract rapidly and for long periods if the soil is deep enough. For example, during a period when little rain fell, the front beneath a stand of tea in Kenya descended to about 4.5 m in 80 d; throughout this time the stand extracted at a very conservative rate of 3.4 mm d^{-1}, or 0.56 E$_o$[23].

An extraction front can also be detected beneath annual stands growing on stored water. The front has characteristics similar to the root front (Chapter 2), moving downwards at a conservative rate provided the soil is moist through the profile and reasonably uniform in structure (Fig. 4.8). As for the root front however, the factors controlling the velocity of the extraction front are not understood. It is slower in clays and loams than in light, sandy soils (where rates up to 70 mm d^{-1} have been measured for a C4 cereal[19]). It sometimes differs between stands in a similar soil (cf. the two groundnuts in Fig. 4.8). In soils of intermediate texture, the velocity of the front ranges from 10–40 mm d^{-1}. In Figure 4.8, it moved faster among the cereals than the legumes, but this is not always so[17]. If the front reaches a dry layer of soil, it might cease descending, only continuing if the soil beneath it is wetted by rain-water percolating down from above[5]. Compacted soil horizons might reduce the velocity of the front or prevent it moving further.

Nevertheless, the close correspondence in annuals between root and extraction fronts implies that water is extracted from a layer of soil as soon as, or very shortly after, the root front reaches it. Subsequently, the rate of extraction increases, presumably as roots proliferate in the layer, and reaches a maximum several days after the front passes through the layer (Fig. 4.9). Finally, the rate decreases to very small values one to two weeks later, mainly because the hydraulic resistance to the flow of water to the roots increases considerably as the soil dries. The depth of fastest extraction therefore lags behind the extraction front. Usually (e.g. Fig. 4.9b) the rate of maximum extraction decreases with depth, much as does l_v. However, the relation between l_v and extraction rate is not usually linear: inflow per unit root length is sometimes greater at the lower root densities[26].

These and other data indicate that the rate of extraction by a root system exploring a store of water is limited for two reasons: first, when roots reach a layer of moist soil, extraction from that layer is slow because there are few roots there (l_v is small); second, by the time l_v in that layer has increased to a value capable of extracting very rapidly, the soil water content (and potential) has decreased and is itself limiting inflow. For the millet crops in India and Niger referred to earlier[19], l_v did not increase much above 0.5 cm cm^{-3} in most of the profile, and may have been prevented from increasing further than that because the soil was too dry.

Therefore, the main factor limiting extraction rate by stands growing on a reserve of soil moisture is the rate at which roots extend and proliferate; and for a crop in these circumstances to extract appreciably faster than 3 mm d^{-1}, extension of roots would need to be fast enough for

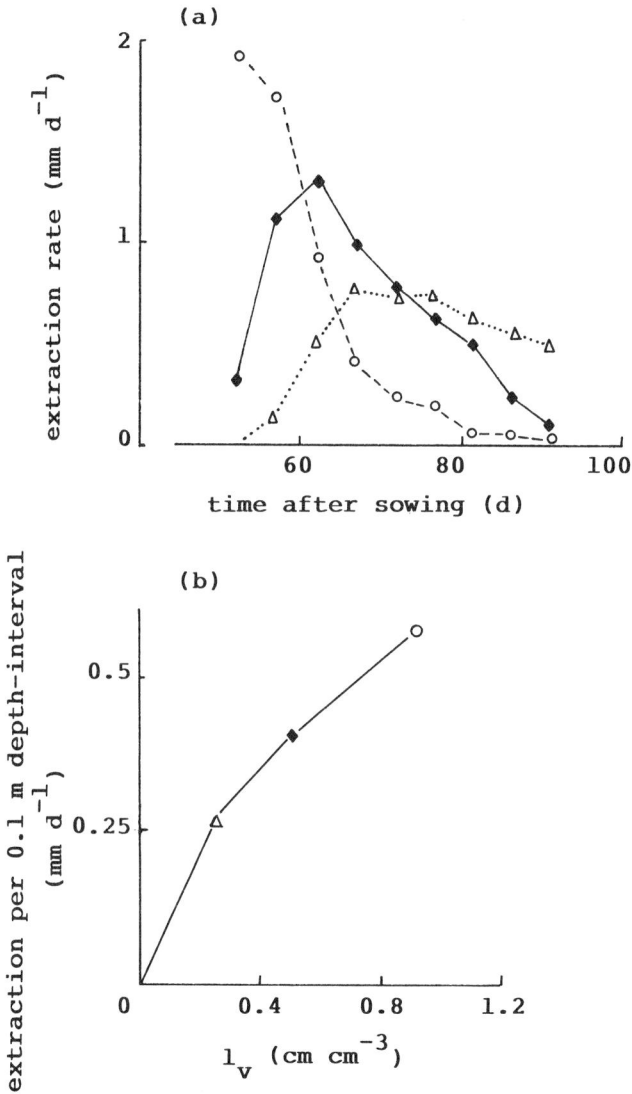

Fig. 4.9. Rooting density and water extraction by stands of groundnut. On a drying soil at Hyderabad, India, for depths of 0 to 0.3 m (○, – – – –), 0.3 to 0.6 m (◆, ——), and 0.6 to 0.9 m (△, ·····): (a) shows change with time in the rate of extraction; (b), the relation between extraction rate and rooting density (l_v) at the time of maximum extraction. Source: Simmonds and Azam-Ali, 1989.

the root front to proceed well ahead of the extraction front, and for a dense root system ($l_v > 1.0$ cm cm^{-3}) to develop in the layer immediately behind the front before water was depleted from that layer at an appreciable rate.

Profiles and amounts of extracted water

A dense root system, such as occurs in the uppermost layers of soil beneath continuous stands, can generally reduce a soil to around its permanent wilting point. Below the uppermost layers, the profile of extracted water is more often similar to the profile of l_v[7] than to that of the permanent wilting point. Accordingly, much less water is usually extracted at depth than is estimated to be 'available'. The difference between extracted and available water depends on the uniformity of the soil, specifically the presence of compacted layers that might restrict root extension, and on the other factors that determine rooting density.

The amount of extracted water therefore depends much on those factors that govern the rate and duration of expansion (Chapter 2), and the following examples illustrate some common patterns. Figure 4.10 shows the minimum soil water contents when stands of pearl millet had ceased extracting during a drought in a deep sand and a moderately deep sandy loam. In the sand (in Niger) the volumetric water content (after drainage) was about 11% and fairly uniform down the profile; the root system reduced the content to about 4% at depths of 0.4–1.6 m, though above 0.4 m evaporation from the soil surface continued to deplete soil water to a volumetric content of less than 2%. The loss from this soil throughout the root profile down to 1.8 m was about 110 mm, of which 20–25 mm was lost directly from the soil surface. In the medium-deep loam, the initial, fully drained, water content of 20–25% was reduced to 5–10% at depths between the surface and 0.5 m. Deeper in the soil where there were very few roots (probably because root extension was restricted by a dense soil layer), the water content was 15% when the stand ceased transpiring. The water lost between the surface and a depth of 1 m was about 90 mm, of which about 20 mm were lost directly from the surface. The amount of water extracted by this stand was only a third of that initially held in the soil, and about 60% of the 'available' water. (Reference 6 gives comparable analyses for clay soils.)

The differences between these examples were caused entirely by soil factors. Where the soil is several metres deep and has no barriers to the root system, differences are more commonly caused by developmental factors. Indeterminate genotypes can usually extract more than determinates. This is so whether the stand is a perennial, having already established a root system, or an indeterminate annual whose root and

volumetric water content (%)

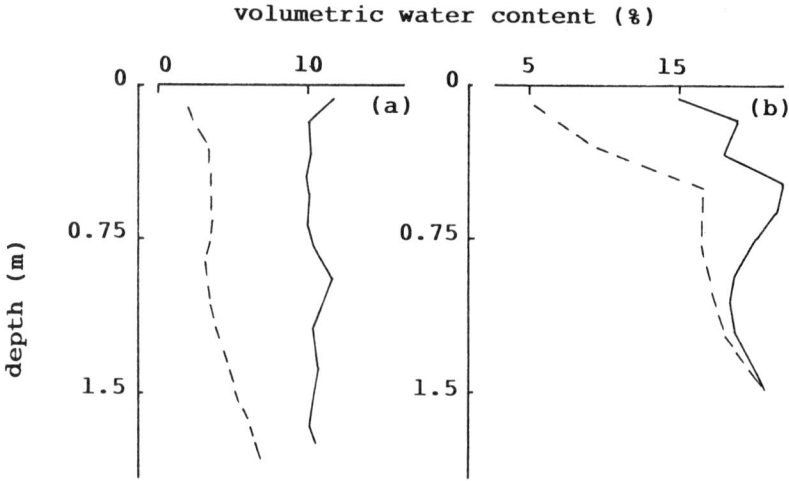

Fig. 4.10. Examples of the water extracted by stands of pearl millet. Showing the volumetric water content when fully drained after irrigation (———) and when extraction by the crop ceased (– – – –) for (a) a sandy soil at Niamey, Niger, from Azam-Ali, Gregory and Monteith, 1984b, and (b) a medium deep alfisol at Hyderabad, India, Squire *et al.*, 1984.

extraction fronts move simultaneously. The tea referred to earlier extracted 200 mm over 80 d, mainly from stored water, and probably could have continued to extract at the same rate for several months, since the soils were 11 m deep[23]; and a stand of cotton in the Yemen (Fig. 4.11) extracted 400 mm of water stored from the surface down to below 3.5 m (extraction front: 3 mm d^{-1} for 110 d).

In contrast, determinate types extract little water after expansion ceases near flowering (Chapter 2). At the same site as the cotton in the example just given (Fig. 4.11), sorghum extracted for a period of 90 d, and maize for 50 d. The maize extracted little water below 1 m, eventually taking only half as much water from storage as the cotton. The root front velocity was slower for the maize (2.8 cm d^{-1}), but the main limitation was probably developmental duration. This effect of development on root extension limits the time for which the front moves, and extracts rapidly, to an even shorter period, 30–40 d, in short-season cereals. As a result of this effect, many cereals extract little water from a store during grain filling.

The available water per unit depth of soil in this example from the Yemen was comparable to that in Figure 4.10b: the fivefold difference in transpired water between the cotton and the millet was the result of

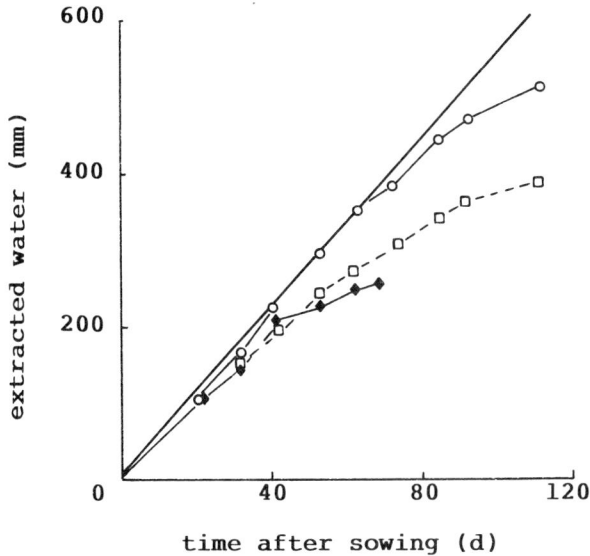

Fig. 4.11. Cumulative water extracted by three subsistence crops growing mainly on a store of water. On a deep soil in the Yemen, irrigated before sowing, and receiving little rain after 20 d: (○) cotton, (□) sorghum, (◆) maize. The continuous line is cumulative E_o. Adapted from Williams, 1979, Figure 24.

differences in determinacy and soil depth. Despite such variation, the profile and amount of extracted water are usually conservative for a specific site and genotype. Soils are therefore sometimes compared in terms of the 'extractable water'[24] rather than available water. However, the extractable water at a site must be qualified by reference to a standard genotype.

Interdependence of root and shoot

This section attempts to draw together information given earlier in this chapter, and in Chapter 2, to show how the canopy and root system are linked in the regulation of transpiration. As the rate at which water is extracted from the soil is equivalent to the rate at which it evaporates from the foliage, the two systems can be related by

$$(g_t L)v^* = RI \tag{4.7}$$

where g_t is a term combining leaf and aerodynamic conductances but dominated in dry conditions by leaf conductance,

L   the leaf area index of the canopy,

v* the leaf-to-air water vapour concentration difference, representative of the canopy as a whole,

R   the total length of root per unit area of ground,

and   I   the mean inflow of water per unit length of root.

Sometimes, it is clear which of the physiological factors in equation 4.7 limit transpiration rate. For example, when the air is humid (v^* is small), g_l is generally large enough not to limit, and E_t is restricted by the rate at which L increases, as determined principally by effects of temperature on development and the expansion of leaves; the inflow, I, is probably conservative since R/L is itself conservative. During a drought, if the root system meets a barrier in the soil, transpiration slows and soon stops because R can no longer increase and inflow decreases.

On many occasions however, it is unclear whether transpiration rate is governed by attributes of the root system or the shoot system. For example, in moist conditions, senescence affects both g_l and I: in a pearl millet crop referred to earlier (Hyderabad[19]), R changed little between flowering and harvest, L decreased slightly, g_l halved, and I decreased to less than half the value at flowering. In dry conditions, each of the pairs, R and L, and I and g_t, respond on different time scales to compensate for change in the other of the pair, but there is often little direct evidence to show which attribute is primarily responsible. The hindrance to progress is the difficulty in the field of manipulating any one of these four physiological variables so as to observe the responses of the other three. Removing leaves of a droughted stand – a drastic recourse – has been found to bring about an increase in the conductance of the remaining leaves, consistent with the conclusion that transpiration is limited by the root system in these circumstances. In the field, a more subtle means of manipulating the balance is through the population density (as described later). The responses at any time also depend on the potential evaporation rate.

The relation with potential transpiration

During wet seasons, clouds reduce the solar radiation reaching the ground, and continuous evaporation on a regional scale brings down the saturation deficit (Reference 24, Chapter 2). The potential evaporation rate is usually $4-6$ mm d^{-1}, and most stands are able to transpire near this rate. Stands irrigated during hot, dry seasons, are exposed to much higher potential rates. A few such stands can still transpire near the potential (for example, the irrigated rice in Figure 4.3) but most others do not transpire as rapidly, even with irrigation. There have been few field investigations into how transpiration is restricted when E_o is very large, but it is suspected that the negative effects of saturation deficit, found in controlled environments, are responsible.

Saturation deficit limits transpiration rate by slowing leaf expansion and reducing leaf conductance. In moist soil, both responses do not always occur; it is not known why this is so, and how the responses are modified by factors of the root system and soil. The effects of D are more pronounced for crops transpiring on a store of water when the supply of water by the roots limits transpiration. A change in D, acting through v^* in equation 4.7, should be compensated by a change in the opposite direction in L or g_l (or both). Observations in different climates suggest this usually happens: i.e. the canopy conductance decreases as E_o increases, whereas the transpiration rate is more stable. For example, stands of pearl millet at Hyderabad, India, where E_o was 5 mm d^{-1}, and Niger where E_o was 8 mm d^{-1}[19] transpired at similar rates, 2–3 mm d^{-1}, when the extraction front was moving, but achieved these with different canopy conductances – 0.8 cm s^{-1} at Hyderabad, 0.4–0.5 cm s^{-1} at Niamey. Consequently, E_t/E_o was larger at Hyderabad (where it was about 0.5) than in Niger (0.3).

Nevertheless, there are indications that stands depending on a store of water transpire slightly faster when D is larger (indicating a possible link between D and either root length or inflow). This response was systematically investigated in controlled environment glasshouses in the UK, using stands of groundnut with roots free to explore the soil[27]. Treatments consisted of different values of v^* (soil conditions being similar at the start of growth). Each of the treatments responded as in Figure 4.7b: i.e. E_t reached a stable value because E_t/f decreased, but the stable value of E_t was slightly larger for stands in drier air.

This result implies that water was supplied to the foliage faster when the air around the foliage was drier. The result could not be explained by effects of the water potential gradient on inflow, but there was circumstantial evidence that R was larger when D was larger (though R was not measured). There was more dry matter in roots in these treatments and both the extraction front velocity and the rate of extraction deeper in the profile were faster (implying l_v increased more rapidly). This experiment shows that the rate of transpiration, even when limited mainly by the root system, is still sensitive to the atmospheric environment. The physiological relations between extraction rate and saturation deficit have hardly been investigated in the field.

Relations with canopy temperature

The relations between the variables in equation 4.7 are further complicated because the value of the vapour concentration difference, v^*, is not independent of the other variables. It is very sensitive to any change of leaf temperature, since this determines the saturation vapour concentration in the sub-stomatal spaces. Of the physiological variables, leaf conductance has the greatest influence on leaf temperature – at least in the short term and on a local scale. When conductance is large enough not to restrict transpiration, the leaf temperature will probably be within 1–2°C of air temperature, and v* will be determined mainly by the saturation deficit of the air. If conductance decreases, less of the solar radiation absorbed by the canopy is used to evaporate water, and leaf temperature rises. (Some of the articles in Reference 1 examine the relations between transpiration and leaf temperature.)

The relation between leaf conductance and leaf temperature depends on other factors such as the incoming solar radiation and the windspeed, and the orientation of leaves. It is difficult to generalize, but foliage of droughted stands will typically be 3–6°C warmer than air temperature when transpiring slowly, and 10–12°C when stomata are closed. Such temperature differences have very great effects on the leaf-to-air vapour pressure or concentration difference. The value of $\varepsilon_w D$ based on total water transpired is still conservative for droughted stands, since they usually transpire most of the season's water when stomata are open and the leaf-to-air temperature difference is small.

The high tissue temperatures that occur during drought probably have more importance for development and survival, but the physiological variables, L and R in equation 4.7, respond so as to raise the R/L ratio and thereby conserve leaf temperature in many instances. These responses are more influential in plants such as pearl millet than groundnut, but are generally more effective at lower populations than those at which stands are normally established in wet seasons (see below).

Modification by cultural factors

There has been little systematic investigation in the dry tropics into the extent to which transpiration can be modified by management. The effect of mixed cropping is still uncertain. Shelterbelts and windbreaks, consisting of rows of tall plants, usually reduce the amount of water transpired (and therefore the dry matter produced) by crops in dry weather. Shade trees also do this, but nevertheless are sometimes beneficial by reducing leaf-scorch of the canopy beneath them[27]. The effects of fertilizer and plant population can be substantial, as in the following examples (each for one species at one site).

Fertilizers

Adding nutrients to soil has several effects on the processes controlling transpired water. Fertilizer sometimes reduces the duration of the crop, but has a much greater effect on the rate of expansion of the root system. For barley in Syria,[7] the stimulation of root length (and leaf area) by adding nutrients caused stands to transpire more rapidly early in development, but more slowly near the end of the life-cycle, since the stands had then used most of the water. The nutrients also increased the amount of extractable water, mainly through an increase in l_v. At depths to 1.2 m, over which l_v was measured, the effect was small (Fig. 4.12); for instance, a doubling of l_v near the surface had little effect on extractable water. The greatest effect appeared to be at depths between 1.2–1.8 m where the fertilizer also seemed to increase l_v. Between the soil surface and 1.8 m, fertilizer increased the extractable water by 24 mm, from 168 to 192 mm.

The effect of fertilizer on transpiration was even smaller at a drier site than that just referred to, and more generally, it seems the effect will

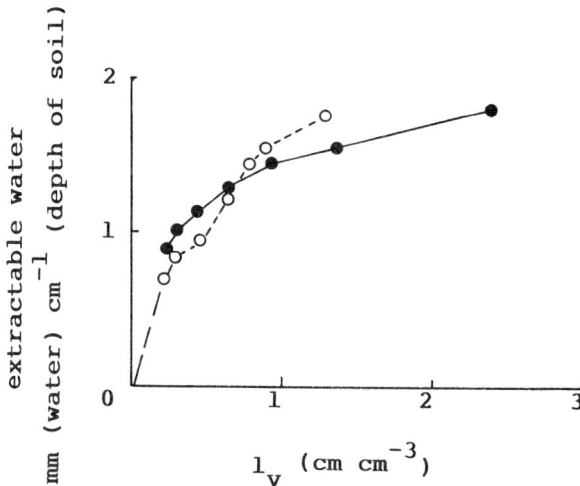

Fig. 4.12. **Effect of fertilizer on root length and extracted water for barley in Syria.** During a period with little rainfall: extractable moisture is the difference between maximum water content after recharge and the content at harvest; rooting density (l_v) is the mean of measurements at anthesis and maturity. Measured in seven 0.15 m depth intervals from 0.15 to 1.2 m (omitting 0 to 0.15 m layer since most water from this lost by E_s); both variables decreasing with depth. Symbols: ○, no fertilizer; ●, optimum fertilizer for the site, 60 kg P_2O_5 ha^{-1}, and a split dressing of a total of 60 kg N ha^{-1}. Between 0.15 and 1.2 m: total root length 7.0 km, extractable water 130 mm for unfertilized; 9.6 km m^{-2} and 139 mm for fertilized. Source: data for local cv. Arabic Abiad, at Jindiress, in Tables 1 and 2 of Cooper, Gregory, Keatinge and Brown, 1987.

depend much on the distribution and quantity of rainfall. The fertilized stand in Figure 4.12 might have transpired much more than the unfertilized had there been more rainfall after the root system had expanded: the stimulation of l_v below the soil surface might have been more effective and the decline of E_t prevented.

Plant population density

The population at which a stand is established has a great effect on the rate at which it extracts water[6,19]. At low populations, at which the roots of individual annuals might not mingle substantially, E_t is proportional to the population. In the example in Table 4.3, the transpiration per plant, over the 50 d of measurements, was little affected by population up to at least 12 m^{-2}; but individuals of other genotypes, extending more rapidly, might affect each other at much lower populations. Up to 12 m^{-2} in the example, both the descent of roots and their proliferation, and the rate of extraction, were greater beneath than between plants or rows of plants, but the population was high enough at 23 m^{-2} to support a two-dimensional front. Nevertheless, at a given distance from the plant, the extraction front descended more rapidly as population increased (Fig. 4.8b). The effect was large even compared with differences in the front velocity between sites and genotypes.

This effect of population controls the time that a stand transpires on a finite store of water (sometimes termed the 'water-time'[29]). For example, groundnut (Table 4.3) on a 1 m deep soil transpired for about 50 d when at a density of 22 m^{-2} after which time it had reached a dense stony layer through which most roots did not penetrate. At this time, the measurements ceased, but if the other stands were assumed eventually to extract the same amount of water from the soil as this dense stand, the projected time for transpiration would be 90 d at 7 m^{-2}, and 18 months at < 1 m^{-2} (not shown in Table 4.3). These effects of density on water-time are again much greater than the maximum differences between species grown at the same site and density – though C4 cereals generally transpire more rapidly than the groundnut in this example, and would use the same store of water more rapidly at any population.

Moreover, plant population has a consistent effect on the equilibrium in equation 4.7. It was shown in Chapter 2 that as the population increases from small values, the ratio of root length to leaf area (R/L) decreases (though it is not known why). In the experiment with a bunched groundnut (Table 4.3), leaf movements did not compensate for this modification of the root/leaf balance, and consequently the root length divided by the fractional interception of solar radiation (R/f) also decreased as population increased (Table 4.3). The equilibrium in equation 4.7 was maintained by leaf conductance decreasing with rise in population, i.e. at a given mean

Table 4.3. Effect of plant population density on characteristics of transpiration by groundnut on a drying soil.

Population (m^{-2})	6.6	12.2	22.9
Leaf area index	0.47	0.80	1.3
Fractional interception (f)	0.18	0.26	0.48
Total root length:			
Unit field area, R $(km\ m^{-2})$	3.5	4.0	5.0
Per plant (km)	0.53	0.35	0.22
Transpiration rate:			
Unit field area, E_t $(mm\ d^{-1})$	1.2	1.9	2.3
Per plant $(kg\ d^{-1})$	0.18	0.17	0.10
E_t/f $(mm\ d^{-1})$	7.0	7.4	4.8
R/f $(km\ m^{-2})$	20.0	15.0	11.0

At Hyderabad, India, between 50 and 95 d after sowing. Leaf area index and root length, mean of measurements on 60, 76 and 90 d; transpiration, from mean of 9 determinations of soil water content; fractional interception of total solar radiation, many diurnal measurements.
Sources: Azam-Ali *et al.*, 1989, Rao *et al.*, 1989, Simmonds and Azam-Ali, 1989, Simmonds and Williams, 1989.

inflow, conductance was smaller at a higher population. The consequence was that the transpired water to intercepted radiation quotient remained high (near 0.4 kg MJ^{-1}), at the lower populations, but decreased at the high population (with probably concomitant raising of leaf temperature and lowering of tissue water potential). These responses illustrate functional effects of the changes in R/L in Table 2.3. The mechanisms controlling the ratio however, are not understood.

Transpiration and dry matter production: synthesis

Many factors of the climate, the soil and the physiology of species influence both the amount of transpired water and the relation between transpiration and dry matter production. The relative importance of the main factors is now shown by examining the differences in dry matter production between selected stands. Finally, some applications of the principles in this chapter are indicated.

Factors limiting productivity

A comparison of sorghum, cowpea and groundnut grown on stored water in the Philippines illustrates some typical effects of photosynthetic

Table 4.4. Factors controlling dry matter production of three rain-watered crops in the Philippines.

Species	Sorghum	Groundnut	Cowpea
Crop duration (d)	78.0	103.0	68.0
Maximum extraction depth (m)	1.0	1.4	1.2
Rate of extraction (mm d^{-1})	2.6	2.7	2.6
Transpired water (mm)	206.0	277.0	181.0
Total dry matter (g m^{-2})	760.0	370.0	280.0
Dry matter/transpired water ratio (g kg^{-1})	3.7	1.3	1.6

Data from Angus *et al.*, 1983.

efficiency and crop duration (Table 4.4). For the C3 species, rooting depth and extraction rate were similar, but the cowpea produced 65% as much dry matter as the groundnut. The difference was caused mainly by the shorter duration of the cowpea – it took less water than the groundnut from a similar store of soil. The C4 species produced much more dry matter than either C3: it extracted water slightly faster, but had a much larger dry matter to transpired water ratio.

The large differences in production between the stands of pearl millet in Table 4.2 show effects of several environmental factors (since the genotype was the same). The total dry matter above ground at final harvest decreased by a factor of 8.5 from the stand grown on irrigated soil in a humid controlled environment to one grown on a drying soil in Niger. (Roots, weighing 30–50 g m^{-2}, made little difference to the range of values but were a much greater proportion of total dry matter in the drier stands.) There was a maximum difference of a factor of about three in each of transpired water and the dry matter to transpired water ratio. For Stand 2 grown with irrigation at Hyderabad, the reduction of transpired water to 0.68 of the value in moist conditions (Stand 1) was caused mainly by effects of senescence on leaf area and leaf conductance. These reduced the rate of transpiration but hardly affected the duration. The water available was not limiting since it was supplied by irrigation. The ratio ε_w was also smaller than in the most humid conditions because the saturation deficit was larger.

For the stands grown on drying soils at Hyderabad and Niamey, transpired water was reduced to 0.32 and 0.36, respectively, of the values for Stand 1, but the cause of this reduction was different at the two sites. At Niamey (Stand 4), the principal cause was that the sandy soil held little water, so even though the roots reached almost 2 m and depleted the soil profile uniformly (Fig. 4.10), they extracted only 80 mm of water. At Hyderabad (Stand 3), there was about twice as much water per unit depth of soil as at Niamey, but the roots descended more slowly, with the result that the volume of soil accessible to the fully expanded root system was smaller than at Niamey. Additionally at Hyderabad, a compacted soil horizon restricted proliferation of roots at depths below 0.5 m, and much water was left in the soil when transpiration ceased. These two stands abstracted similar amounts of water, but the air was so much drier at Niamey that the ratio ε_w was half the value it was at Hyderabad.

These comparisons, and other information in this chapter, illustrate that in different circumstances, dry matter production can be limited by any of the following factors:

- the rate of transpiration, as affected by
    - restrictions of the canopy (e.g. senescence);
    - expansion of the root system (soil physical properties);
- the duration of transpiration, as affected by
    - developmental timing;
    - soil depth and water holding capacity;
- the dry matter to transpired water ratio, determined by
    - photosynthetic efficiency (the C3, C4 difference);
    - saturation deficit;
    - and other constraints (low water potential, nutrient shortage).

Applications

Efficient use of water

In dry regions, agricultural production can be assessed in terms of a water-use 'efficiency', defined variously, but commonly as the amount of dry matter produced per unit water evaporated, from all surfaces, during a specified time. In dry regions, the amount of water lost is usually similar to the amount of rainfall, less any surface runoff or drainage, but will include any available water stored in the rooting zone at the beginning of the season. The efficiency is therefore influenced by factors of the plant – particularly the ratio ε_w – and of the soil and atmosphere. It is difficult to increase ε_w of a crop by husbandry, but it is sometimes possible to take advantage of the inverse relation between ε_w and the saturation deficit, by growing crops in a humid period (when the transpired water will yield a greater return of dry matter than when the air is drier.) Generally, reducing runoff and evaporation from the soil surface effectively raises the water-use efficiency.

These losses can be reduced in two ways. First, if the rainfall can be conserved and then concentrated as near as possible to the plants, more of it will penetrate deeper in the soil, and less will evaporate from the surface. A greater efficiency was gained in this way in the lowveld of

Zimbabwe, using a system of tied furrows, which increased yields of cotton, sorghum and maize by a factor of 1.2–1.4, depending on soil type and cultivar[30]. Second, if the canopy and root system can be made to expand rapidly, for example by applying fertilizer or raising the plant population, evaporation from the soil surface will be restricted by the shade from the foliage, and more water will be extracted by roots from the surface layers of soil. These effects probably caused the differences shown in Figure 4.1b and c. In Syria, applying fertilizer increased the water-use efficiency of barley by a factor of 1.3–1.6, depending on cultivar and site[7]. (A possible contributory factor to the greater efficiency – not confirmed – was an increase in ε_w in response to fertilizer.) In Botswana, increasing the population from 1.3 m^{-2} to 5 m^{-2} and higher increased the water-use efficiency by a factor of 1.7[5] – though the water was used more rapidly, with important effects on yield (see Chapter 5).

Irrigation

When a soil is around field capacity, a stand transpires at a rate determined by the demand for water by the foliage (the demand-limited rate). If the water this stand extracts is not replaced, its transpiration rate will, in time, decrease to a rate determined by the extent and density of the root system, and the soil water potential in the rooting zone (the supply-limited rate). Since dry matter production is closely related to transpiration, it should be about maximum for the climate when the stand transpires at the demand-limited rate, but less than maximum otherwise.

To achieve the greatest possible dry matter production, irrigation should be given to keep transpiration at the demand-limited rate, but to achieve the most efficient use of water, irrigation should be given as infrequently as possible to minimize evaporative losses from the soil surface. The effective management of irrigation requires knowledge of the following characteristics of climate, crop and soil.

First, the value of the demand-limited rate, which is commonly determined as the potential evaporation rate multiplied by a 'crop factor'. This factor is sometimes determined by measurement, but is more often a value, judged typical of the specific crop and soil. Demand might be estimated more realistically by dividing the maximum rate of dry matter production for a genotype by an appropriate dry matter to transpired water ratio, ε_w. If production is estimated as the product of the daily solar radiation, S, fractional interception, f and the conversion ratio, ε_s (Chapter 3), the demand on any day is $Sf\varepsilon_s/\varepsilon_w$[29]. Maximum values of these attributes are known for the main species, but it is not certain how useful this approach might be, since f and ε_s are both reduced by dry air independently of the amount of water in the soil.

Another factor is the period of time following irrigation for which a stand will transpire at the demand-limited rate. For a stand with a static root system, the change between demand-limited and supply-limited transpiration occurs when a certain amount of water has been abstracted from the root zone (i.e. when a certain soil water deficit has accumulated). The period at the demand-limited rate can therefore be estimated as the limiting soil water deficit divided by the mean rate of evaporation. If a root system is expanding down into a soil, the limiting deficit increases in time; and might eventually decrease if part of a root system becomes less effective in abstracting water, as a result of senescence for example.

A third factor is the amount of water, applied at each irrigation, that will be enough to bring the water in the root zone to near field capacity, but not so much that some water will drain beyond the reach of the roots. This amount depends on the depth reached by the root system and the profile of root length density. For example, Figure 4.13 shows changes in the water content of a soil during three cycles of irrigation and drying. The amount of water abstracted by the maize changed little with time, since its root system ceased expanding soon after the first irrigation. The amount taken by the cotton increased with time as the roots penetrated deep into the soil. If all the soil beneath the maize had been irrigated to field capacity each time, much of the applied water would have drained through the profile.

Lastly, when there is little water available for irrigation, the time of application might have a strong influence on the yield. The returns of dry matter will be greatest if irrigation can be given during the most humid part of the season, when ε_w will be largest. To achieve the greatest reproductive yield however, irrigation might have to be given at a certain stage in development. In some determinate cereals, the critical stage is that during which the reproductive sink is determined (during and after flowering). In determinate legumes that are unable to retranslocate dry matter from vegetative structures to the reproductive sink, the yield will be much greater from irrigation given after, rather than before, reproduction begins.

Concluding remarks

The method of analysing dry matter used in this chapter has the advantage over that in Chapter 3 in that the term, $\varepsilon_w D$, is much more conservative than the conversion ratio, ε_s. It has the disadvantage that measurements of transpired water take more time and are more technically demanding than those of intercepted solar radiation. Moreover there are still uncertainties. When there is much water in the upper layers of soil, it is unclear how dry air restricts transpiration by reducing leaf conductance and area; and when a root system is forced to extend into lower layers of

soil water content (volumetric percentage)

(a) maize

soil water content (volumetric percentage)

(b) cotton

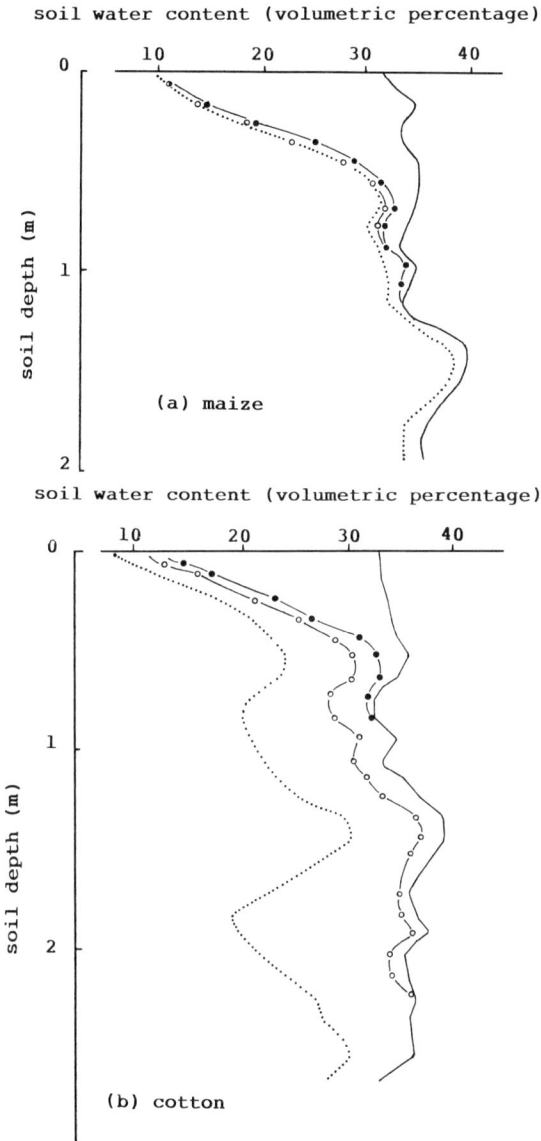

Fig. 4.13. Soil water content profiles showing water extracted by crops over three irrigation cycles. For (a) maize and (b) cotton, in the Yemen. The soil was wetted to the approximate field capacity (——) after each irrigation. During the first drying cycle, sowing to day 28 after sowing, the stands dried the soil to the water content indicated by the symbol ●; during the second, day 29 to day 56 (26 d), to the content indicated by ○; and during the third (a further 50 d for maize, 90 for cotton) to the line ⋯⋯. Source: Williams, 1979, Figure 18.

soil to obtain water, it is unclear what governs the rates of extension of the various rooting elements, and the inflow of water per unit root length, and how drought and nutrients affect the dry matter to transpired water ratio. These uncertainties in the physiology prevent a thorough analysis of productivity in that part of the tropics where yields are smallest and most unreliable.

References

1 Penman, 1948, Thom and Oliver, 1977, Monteith, 1981. The article by K. A. Edwards and J. R. Blackie (1979) pp. 18–22 in Reference 2 is an ideal (brief) introduction to the subject. For practice, see Doorenbos and Pruitt, 1975 (revised 1977), and Frere and Popov, 1979, and commentary by Batchelor, 1984. For applications in research, see References 4, 5, 9 and 10 this chapter.
2 The 1979 Special Issue of the *East African Agricultural and Forestry Journal*, Vol. **43**, titled 'Hydrological Research in East Africa', gives techniques, results and analysis of the components of evaporation, and other factors in the hydrological balance of tea, forestry, rangeland and cultivated land in Kenya. Data in this issue were used by Squire and Callander (1981) to derive the idealized hydrological balance for tea in Table 4.1. Another Special Issue of the same journal, Vol. **44**, 1984, titled 'Dryland Farming Research in Kenya' has several papers on evaporation in semi-arid areas of Kenya.
3 Example of micrometeorological technique; Callander and Woodhead, 1981 (tea, Kenya). See also Reference 13.
4 Williams, 1979, for several crops (mainly cotton, sorghum, maize and pearl millet) in the Yemen, including some measurements in farmers' fields.
5 Rees, 1986c, sorghum in Botswana at different populations (weather dry).
6 Population, groundnut; Simmonds and Williams, 1989 (E_s, E_t); Azam-Ali *et al.*, 1989 (ε_w, f, E_t/S_i); Simmonds and Azam-Ali, 1989 (E_t/f; l_v/f).
7 Barley, nutrients, Syria; Cooper, Gregory, Keatinge and Brown, 1987, Gregory, Shepherd and Cooper, 1984.
8 Sardar Singh and Russell, 1980 (maize/pigeon pea intercrop in wet season, then pigeon pea in dry season, at Hyderabad, India).
9 Kowal and Kassam, 1973 (maize), Kassam and Kowal, 1975 (millet), Kassam, Kowal and Harkness, 1975 (groundnut), all at Samaru, Nigeria. See also Mugah, J. O., Lenga, F. K. and Stewart, J. I., pp. 88–95 in the 1984 *EAAFJ* Special Issue (Reference 2).
10 Batchelor and Roberts, 1983 (inundated rice, Sri Lanka).
11 Soil surface evaporation; Black, Gardner and Thurtell, 1969, Ritchie, 1972, Tanner and Jury, 1976.
12 Transpiration by porometry; Azam-Ali, 1984 (groundnut in India, dry), Wallace, Batchelor, Dabeesing and Soopramanien, 1990 (sugar cane, maize intercrop, Mauritius). Also Reference 10.
13 Rijks, 1971, 1976, and his earlier papers on evaporation from cotton in the Sudan and south Arabia.
14 For example Matthews *et al.*, 1988a, also Reference 6.
15 Dry matter, transpired water (principles with examples); Bierhuizen and Slatyer, 1965 (mainly temperate), Monteith, 1986 (general), Tanner and Sinclair, 1983 (mainly temperate), Cooper, Gregory, Tully and Harris, 1987 (dryland: west Asia, north Africa). See also Further reading, this chapter.

16 Photosynthesis/transpiration ratio; El-Sharkawy and Cock, 1984 (cassava).
17 Angus *et al.*, 1983.
18 Azam-Ali, 1984, also Reference 10, and Jarvis *et al.* in Further reading, below.
19 Pearl millet; Squire *et al.*, 1984 (Hyderabad), Azam-Ali, Gregory and Monteith, 1984b (Niger, plant population).
20 El-Sharkawy, Cock and Held, 1984, Wien, Littleton and Ayanaba, 1979.
21 Connor and Palta, 1981 (cassava, Columbia).
22 Matthews *et al.*, 1988a, Azam-Ali *et al.*, 1989.
23 Cooper, J. D. (pp. 102–21 in 1979 *EAAFS* Special Issue, Reference 2).
24 For further explanation and references; pp. 352–8 in *Russell's Soil Conditions and Plant Growth*, 11th edition (General reading).
25 *Russell's Soil Conditions and Plant Growth*, 11th edition, pp. 358–65.
26 Uptake and l_v: Simonds and Azam-Ali, 1989, Mugah, J. O., pp. 80–7 in 1984 *EAAFJ* Special Issue (see Reference 2).
27 Ong, Simmonds and Matthews, 1987, Simmonds and Ong, 1987.
28 Studies with tea show the many problems of interpreting responses to shade and shelter; McCulloch *et al.*, 1965, Ripley, 1967, Hadfield, 1974, Carr, 1985.
29 Monteith, 1986b.
30 Jones, Nyamudesa and Busangavanye (in press).

Further reading

Rutter gives good general coverage on the hydrological balance (in addition to the papers in Chapter 4, Reference 2), whereas the book edited by Kozlowski contains information on hydrological processes for different types of vegetation, including some agricultural crops. The physics of evaporation is covered in Shuttleworth's report and the chapter by Jarvis, Edwards and Talbot. The 1958 article by de Wit on the relation between dry matter and transpired water, is one of the classics; Stanhill's review includes general and meteorological topics, while Schulze's deals with physiology. For water extraction and soil conditions, see the relevant sections in *Russell's Soil Conditions and Plant Growth* (General reading). The practical significance of the responses described in this chapter is discussed in the review by Cooper, Gregory, Tully and Harris (Reference 15). There are also relevant chapters in the proceedings edited by Taylor, Jordan and Sinclair.

Jarvis, P. G., Edwards, W. R. N. and Talbot, H. (1981) Models of plant and crop water use. In: Rose, D. A. and Charles-Edwards, D. A. (eds) *Mathematics and Plant Physiology*. Academic Press, London, pp. 151–93.
Kozlowski, T. T. (ed.) (1981) *Water Deficits and Plant Growth, VI: Woody Plant Communities*. Academic Press, New York.
Rutter, A. J. (1975) The hydrological cycle in vegetation. In: Monteith, J. L. (ed.) *Vegetation and the Atmosphere, Volume 1*. Academic Press, London, pp. 111–54.
Schulze, E. -D. (1986) Carbon dioxide and water vapour exchange in response to drought in the atmosphere and in the soil. *Annual Review of Plant Physiology*, **37**, 247–74.

Shuttleworth, W. J. (1979) *Evaporation*. Institute of Hydrology, Wallingford, Oxon OX10 8BB, UK.

Stanhill, G. (1986) Water use efficiency. *Advances in Agronomy*, **39**, 53–85.

Taylor, H. M., Jordan, W. R. and Sinclair, T. R. (1983) (eds) *Limitations to Efficient Water Use in Crop Production*. American Society of Agronomy, Madison, Wisconsin, USA.

de Wit, C. T. (1958) Transpiration and crop yields. *Verslagen van Landbouwkundige Onderzoekingen* **64** (**6**): Institute of Biological and Chemical Research on Field Crops and Herbage, Wageningen.

Chapter Five

Partition of Assimilate

Partition of dry matter between the structures of a plant is influenced by all the main environmental and cultural factors. It is intrinsically very complex, depending on the source of assimilate produced by the foliage, the different 'sinks' for assimilate, as determined by the number and size of sub-units in a structure, and on the capacity for movement of assimilate between source and sink, sometimes by way of a transitory store. This chapter considers some general features of partition including the important role of population density. It also shows the influence of the main environmental factors – temperature, photoperiod, drought and nutrients – and examines the relations between number, size and dry mass, mainly for reproductive and fruiting structures.

General considerations: the stand and the individual

The growth in dry mass of a whole plant and one of its structures are represented in Figure 5.1. Part (a) shows the growth of both in relation to time, and part (b), the growth in the mass of the structure in relation to that of the whole plant. The change with time in the dry mass of the structure can be represented by a logistic, or similar, function (Chapter 2, Reference 1) or more simply by assuming the mass increases linearly with time, or with the mass of the whole plant, during most of growth. If linearity is assumed (and the assumption is usually reasonable for plots such as in Figure 5.1b), the effective beginning and end of growth can be determined by linear regression (ideally with a large final sample as a reference). The dry mass of the plant when the structure effectively begins to grow is identified as W_o. During the period of growth, t_w, the change in dry mass can be represented as the corresponding rate of growth by the whole plant (Γ) multiplied by a partition factor, p (the slope of the line in Figure 5.1b). The maximum mass of the structure, $W_s(max)$ is therefore given by

$$W_s(max) = p\Gamma\, t_w \qquad (5.1)$$

143

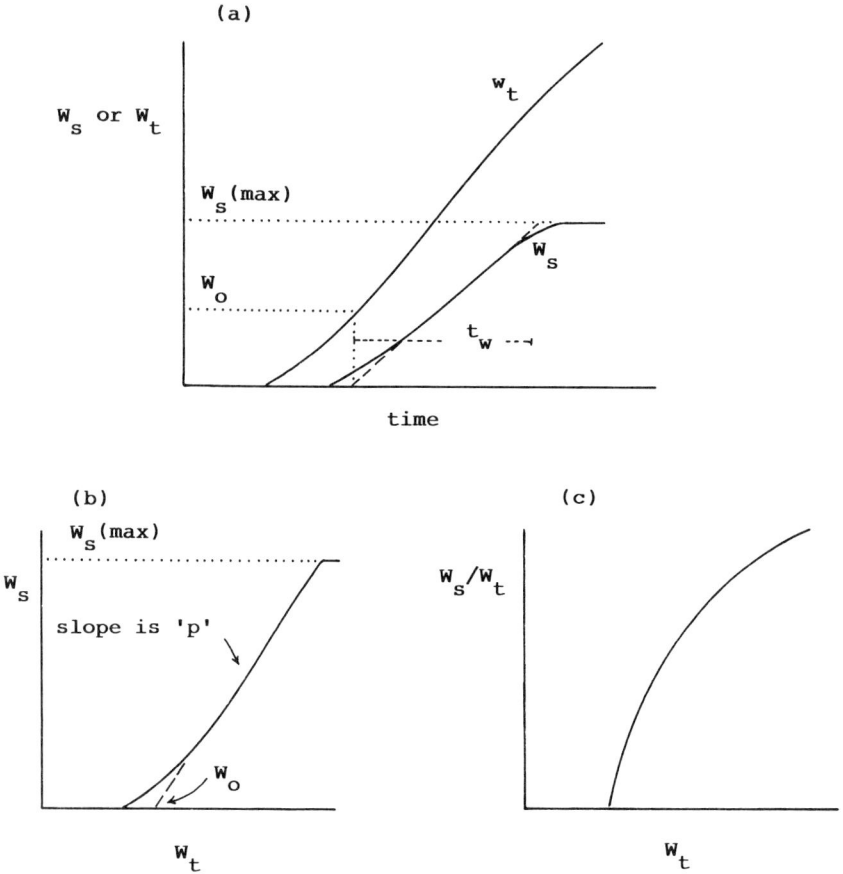

(a)

W_s or W_t

W_s(max)

W_o

W_t

W_s

t_w

time

(b)

W_s

W_s(max)

slope is 'p'

W_o

W_t

(c)

W_s/W_t

W_t

Fig. 5.1. Representation of the growth in dry weight of a structure and of the whole plant. See text for explanation. Symbols: W_t, the cumulative dry weight of the whole plant; W_s, the cumulative dry weight of the structure; W_s(max), the maximum dry weight of the structure; W_o, the value of W_t at the effective beginning of growth in weight of the structure; t_w, the effective duration of growth in weight of the structure; p, the partition factor.

During t_w, the mass of the structure, W_s, is related to the mass of the whole plant (W_t) by

$$W_s = p(W_t - W_o) \qquad (5.2)$$

and the ratio of the mass of the structure to that of the whole plant, W_s/W_t therefore changes with increase of W_t as in

$$W_s/W_t = p(1 - W_o/W_t) \qquad (5.3)$$

as in Figure 5.1c[1]. These simple relations can be used to examine the growth of most structures, though the relation between W_s and W_t is not always linear.

The attributes W_o and p sometimes differ consistently between genotypes. For example, W_o for reproductive structures is smaller in early cereals and legumes than in lates, whereas p tends to be independent of earliness (e.g. Fig. 5.2). In indeterminate types, p for reproductive structures, or root tubers, is usually larger for genotypes whose shoot system is vigorous and much branched than for those with weakly branching shoots (as in cassava, Reference 1). Even in equable environments, W_o is a variable attribute – since it depends very much on conditions, such as solar radiation and plant population, that influence the rate of dry matter production; p is more conservative.

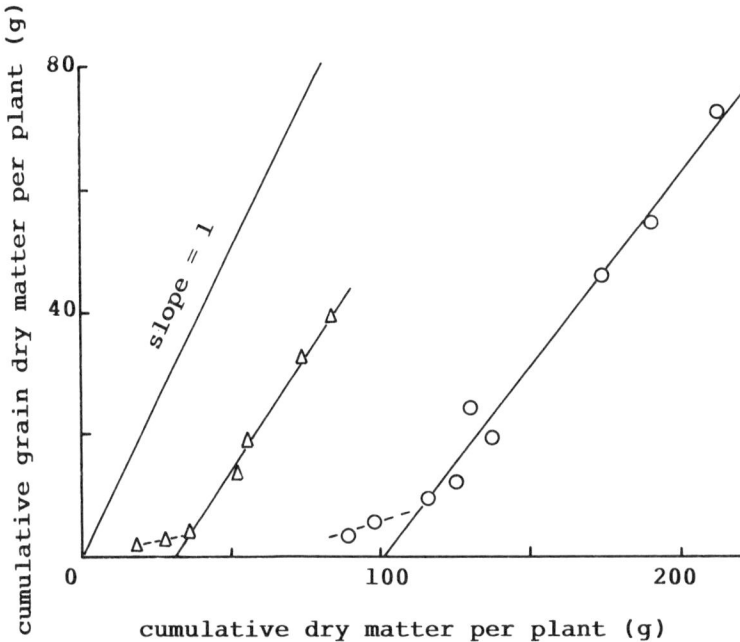

Fig. 5.2. **Partition of dry matter in two cultivars of pigeon pea.** An early ($\triangle$, T-21) and a late ($\circ$, ST-1), differing in the durations of both the vegetative and reproductive phases. For T-21, $W_o = 30$ g, p = 0.74; for ST-1, $W_o = 100$ g, p = 0.62. Original data of Sheldrake and Narayanan, 1979.

The effects of population density

Plant population density (N_p) has important effects on the attributes controlling partition of dry matter. When dry matter is expressed per unit field area, both the total dry matter at harvest, and W_o for most structures,

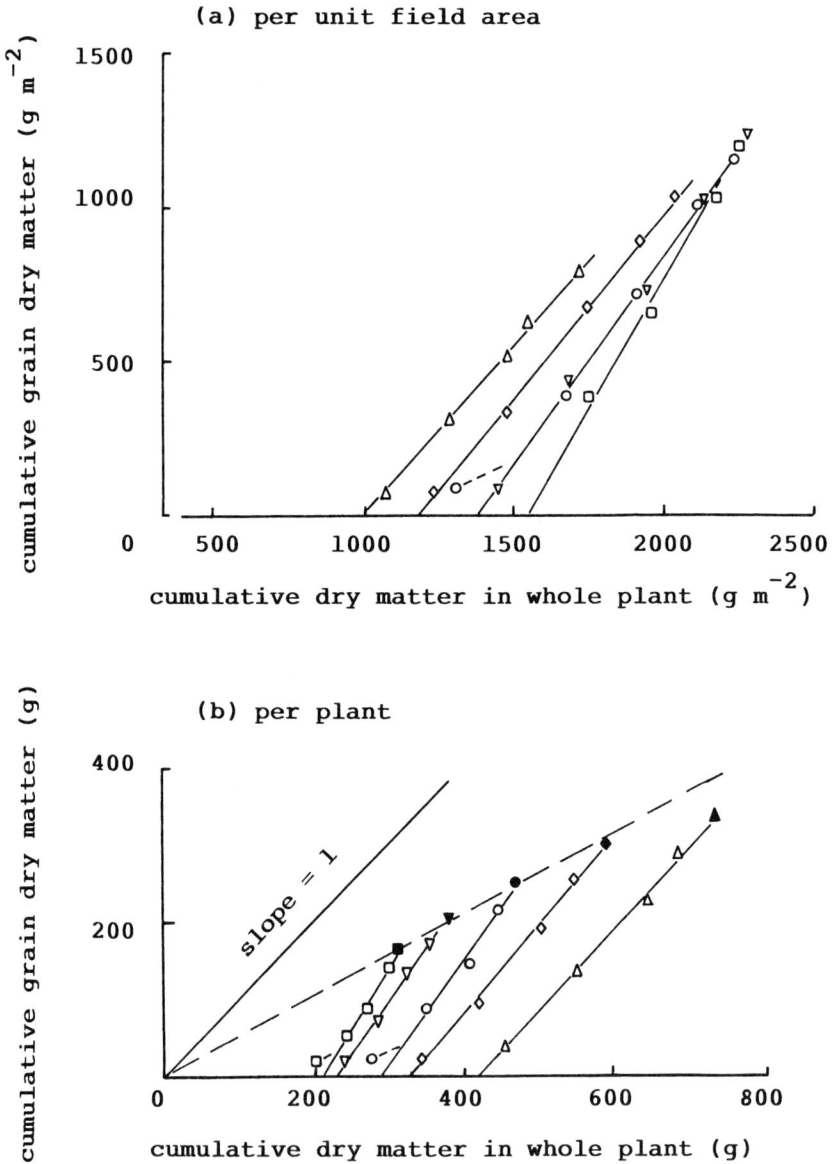

(a) per unit field area

y-axis: cumulative grain dry matter (g m^{-2})

x-axis: cumulative dry matter in whole plant (g m^{-2})

(b) per plant

y-axis: cumulative grain dry matter (g)

slope = 1

x-axis: cumulative dry matter in whole plant (g)

Fig. 5.3. Effect of plant population density on partition of dry matter to grain in maize. In Zimbabwe, at populations (m^{-2}): △ 2.3, ◊ 3.5, ▽ 4.8, O 6.1, □ 7.4. Same data expressed (a) per unit field area, and (b) per plant. Slopes (p) from the highest to the lowest density are: 1.67, 1.34, 1.36, 1.22, 1.07. The dashed line in (b) is the regression of the values at maximum mass (closed symbols) at each population (excluding the stand at 2.3 m^{-2}): y =13+0.5x, r^2 = 0.99, i.e. W$_v$ about zero, and p′ = 0.5. Original data of Allison, 1969.

increase with rise in N_p. The form of the relation is similar to that between leaf area index and fractional interception of solar radiation. When expressed on the basis of a single plant however, the total dry matter and W_o decrease as population rises. At a given population, p should be the same, however dry matter is expressed.

These responses are shown by the analysis in Figure 5.3 of the dry matter in grain and the whole plant of maize grown in Zimbabwe. When dry mass is expressed per unit field area (Fig. 5.3a), W_o increased as population rose up to 4.8 m $^{-2}$, but then changed little (since the additional leaf at higher population intercepted little more solar radiation). When expressed per plant (Fig. 5.3b), grain dry matter, total dry matter, and W_o, all decreased with rise in population. The slope p – the same at a given treatment in both parts of the figure – increased with rise in N_p. (The values of p > 1 in Figure 5.3 imply assimilate was moved to grain from other structures to supplement current dry matter production.)

The changes of W_o and p in such graphs reveal much about the gross physiological responses of a genotype, particularly the extent to which it conserves the proportion of dry matter allocated to each of its structures. This proportion depends on: (i) the total dry matter accrued during filling of the structure as a fraction of that before; and (ii) the partition factor for the structure while it is filling. In Figure 5.3, the proportion of the total mass accrued after grains began filling decreased slightly as population increased. (This was because denser stands senesced slightly more rapidly.) At the same time, however, p for grain increased with rise in N_p from 1.07 to 1.38. As a result, grain mass as a fraction of total mass (the harvest index) hardly changed between 3.5 and 7.4 m^{-2}, such that a regression of grain on total for these treatments passes very close to the origin (the dashed line in Figure 5.3, through the closed symbols). The harvest index for the lowest population was slightly smaller than for the others: the vegetative structures of these large plants were allocated relatively more of the total assimilate, possibly indicating that the sinks in the cob were full.

There are few other sets of data for tropical species consisting of sequential measurements of dry matter over a range of population; only the dry matter at final harvest is usually measured. Therefore it is uncertain whether p generally responds to N_p as in Figure 5.3. Nevertheless, the effect of N_p on the dry matter in fruiting or storage organs as a fraction of that in the whole plant at final harvest still reveals important characteristics of a genotype. Analyses of several population experiments suggest the very conservative harvest index in Figure 5.3b is an exception. Usually, a regression of the final dry mass in fruiting or storage organs on that of the whole plant is still linear, but intercepts the horizontal axis at a positive value. This intercept, W_v, can be defined as the apparent minimum vegetative mass that must be accrued before assimilate can be allocated to reproductive structures. The implication is that, when population is so great that total mass per plant equals W_v, there will be no reproductive yield (but the range of population examined in experiments in moist conditions is not usually high enough to reveal effects when total mass is close to W_v). The slope of the regression of reproductive on total mass at final harvest is identified by p', to distinguish it from the partition factor, p. In Figure 5.3b, p' is the slope of the dashed line through the closed symbols.

Genotypic differences in the response to population

The values of the intercept, W_v, and of the slope, p', differ greatly between genotypes. Contrasting examples are given in Figure 5.4. (The co-ordinates are dry matter at harvest, and are comparable with the closed symbols in Figure 5.3b.) For oil palm (Fig. 5.4a), total mass decreases from 240 to 110 kg per palm as N_p increases. The dry matter produced in fruits decreases by a similar amount, while the dry matter produced in vegetative structures hardly changes; the harvest index falls from 0.63 to 0.27 (Fig. 5.4c). The value of W_v is 80 kg per plant. For pigeon pea (Fig. 5.4b), total dry mass decreases about tenfold from 258 to 25 g per plant, and vegetative and reproductive mass are reduced about proportionately: the harvest index (seed/total) changes little, from 0.35 to 0.29, but if total mass were to decrease further, the harvest index would eventually fall to much smaller values. The intercept, W_v, is about 3 g.

These are two extreme responses: Table 5.1 gives further examples of W_v and the slope, p' determined from data at final harvest in population experiments. Those genotypes with a very small W_v – rice, early pigeon pea, pearl millet – are still able to produce a yield when the mean mass of the individual becomes very small. Those with a large W_v – oil palm, cassava, late pigeon pea – allocate dry matter preferentially to leaves and stems, at the expense of reproduction and fruiting[3].

The values of W_v and the slope, p', are far more conservative than W_o and the partition factor, p. They also have a strong influence on two other attributes of a genotype: the population density at which there is most yield in reproductive or storage organs (considered in Chapter 6); and the stability of the harvest index in the face of change in population. The change in harvest index in relation to change in total dry matter per plant probably takes the same form in most species: the index rises from zero as total dry matter rises above W_v (much as W_s/W_t rises above W_o in Figure 5.1c), eventually reaching a more or less stable value. Stability is achieved at a much smaller plant mass (and therefore is maintained over a wider range of population) in genotypes, such as pigeon pea, with a small W_v, compared with those, such as oil palm, with a large W_v.

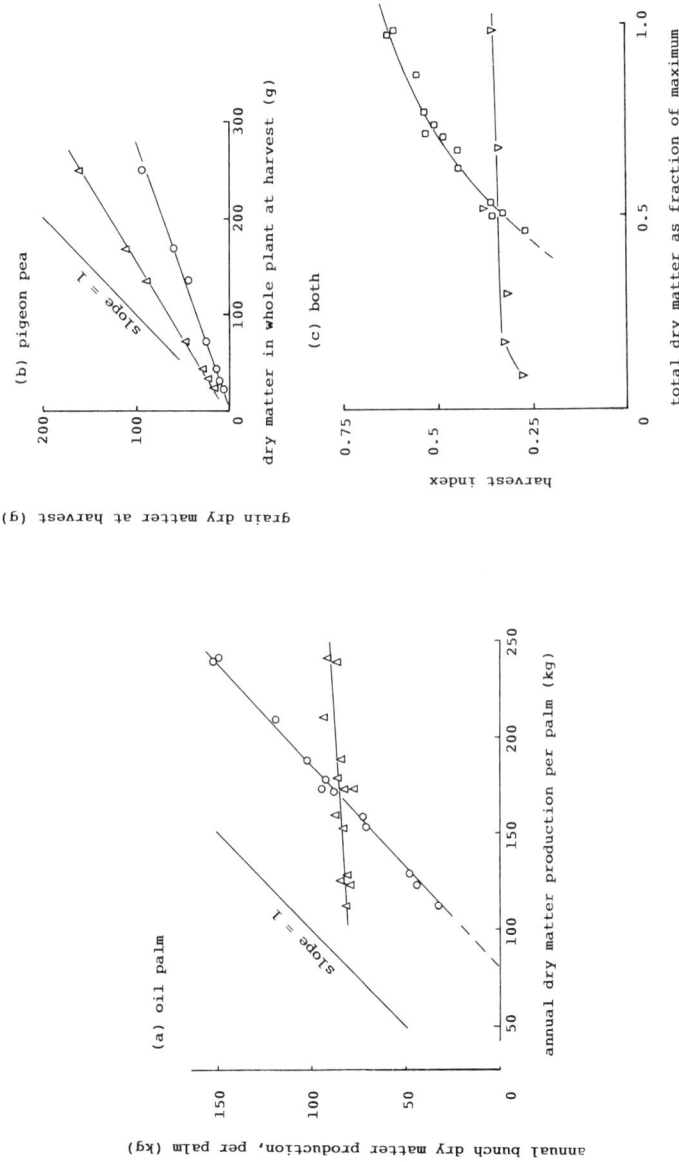

Fig. 5.4. Partition between vegetative and reproductive structures in oil palm and pigeon pea as influenced by dry matter per plant. Reductions in dry matter per plant were obtained by increasing the population density. In (a), $\triangle$ is vegetative dry matter, and $\circ$ is fruit bunch dry matter; both fruit and total dry matter corrected to account for greater energy of oil in fruits; $W_v = 80$ kg, slope p' for fruits is 0.93; original data of Corley, 1973. In (b), $\triangle$ is vegetative dry matter, and $\circ$ pod dry matter (not corrected); $W_v = 8$ g, p' for grain is 0.36; original data of Venkataratnam and Green, 1979. In (c), the harvest index is dry matter in fruits or pods as a fraction of total above-ground dry matter; the horizontal axis is the total dry matter as a fraction of that for the maximum in each experiment: $\square$, oil palm; $\triangledown$, pigeon pea.

Table 5.1. Plasticity of the individual, as indicated by the apparent minimum vegetative mass (W_v) and the ratio (p') of reproductive/total dry mass.

Species	W_v (g)	p' (g g^{-1})	Source of data
Rice	1.1	0.47	Pyare Lal *et al.*, 1982
Pigeon pea	2.7	0.34	Venkataratnam and Green, 1979
Rice	3.2	0.48	Akita, 1982
Pigeon pea	3.8	0.27	Rowden *et al.*, 1981
Pearl millet	4.2	0.55	Carberry *et al.*, 1985
Groundnut	12.0	0.57	Bell *et al.*, 1987
Maize	16.0	0.48	DeLoughery and Crookson, 1979
Cassava	170.0	0.76	Cock *et al.*, 1977
Pigeon pea	196.0	0.19	Akinola and Whiteman, 1974
Cassava	350.0	0.62	Cock *et al.*, 1977
Cassava	1120.0	0.51	Enyi, 1972a
Oil palm	8×10^4	0.93	Corley, 1973.

From experiments on population density, in which the reproductive dry mass per plant at harvest (W_r) is related to the corresponding total dry mass (W_t) by

$$W_r = p'(W_t - W_v)$$

derived from linear regressions of W_r on W_t, where p' is the regression coefficient; r^2 for all regressions was > 0.99. W_r for cassava is the dry mass of root tubers.

The patterns of partition described here – particularly the linear relation between the mass of a structure and that of the whole plant – have implications other than for effects of population density. Little is known of the effect of change in solar radiation on partition, but it is expected that this response would be similar to that to population – since both alter total dry matter. Shading experiments are consistent with this conclusion (Chapter 3, Reference 16). For example, shading groundnut during reproductive growth, which reduced total dry matter to 70% of an unshaded control, had a negligible effect on p for pods, which is expected, since groundnut has a plastic response. Artificially shading cassava caused a greater reduction of p for tubers than for shoots, especially in young plants whose shoots were rapidly extending, which is also expected from the large W_v and the priority given to shoots.

Moreover, a genotype with a very plastic vegetative dry mass has the advantage that some reproductive yield is produced in adverse environments when total dry matter production is very small, but has the disadvantage that in favourable environments, when total production is large, the vegetative organs will still claim a large proportion of the dry matter. In contrast, a genotype with a fixed vegetative dry mass has the disadvantage that it might yield nothing in adverse environments, but the advantage that any dry matter produced in addition to that required for the vegetative structures will mostly be allocated to yield. Some such genotypic differences have their basis in structural or developmental attributes. A cereal, for example, produces all its fruiting units more or less at the same time and clumped in the panicle; whatever the total production, the stem must contain at least enough structural material to support the panicle until the grains mature. In contrast, the oil palm has a much stronger stem, and moreover produces its fruits asynchronously, and so can allocate much dry matter to yield in favourable environments without collapsing. (Some individuals of this species can allocate as much as 70% of the total carbohydrate equivalent dry matter to oil.)

Effects of temperature, photoperiod, drought and nutrients

The physical environment affects partition of dry matter by influencing the partition factors and the periods in which structures increase in mass. Many effects on partition factors are associated, either directly or indirectly, with change in the sink for dry matter in one or more structures. The sink depends on two sets of processes: those governing the potential number of primordia that are capable of growing into structures, such as leaves, stem internodes and grains; and those governing the actual number that survive and accrue dry matter. As in previous chapters, the direction of the response to temperature depends on whether processes are determinate or indeterminate.

Temperature: determinate

The period for which a structure increases in mass is usually related to an underlying process of development. The strictly determinate periods, such as the filling of reproductive structures (cereal panicles, fruits) usually respond as a developmental duration, as in

$$1/t_w = (T - T_{bw})/\hat{\Theta}_w \qquad (5.4)$$

where $\hat{\Theta}_w$ is the integral of time and temperature above the base required for growth to be completed[6]. Other processes will be governed similarly if their duration is, in effect, determinate. For all these processes, a rise in temperature reduces the period from sowing to the start of growth of the structure, and the period of the growth itself. The ratio of t_w for two structures is little affected by temperature as in the general determinate case discussed in Chapter 1.

The determinate sink

In determinate structures, the potential size of the sink is controlled in a way similar to that for leaf size described in Chapter 2. The reciprocal of the time (t_i) required for initiation of a primordium responds to temperature, T, as in

$$1/t_i = (T-T_{bi})/\hat{\Theta}_{1i} \qquad (5.5)$$

where T_{bi} and $\hat{\Theta}_i$ are the corresponding base temperature and thermal duration. The period for which a series of primordia are initiated, t_s, also depends mainly on temperature as in

$$1/t_s = (T-T_{bs})/\hat{\Theta}_{1s} \qquad (5.6)$$

where T_{bs} and $\hat{\Theta}_s$ are the base temperature and thermal duration for initiation of a series of primordia, respectively. The potential number of primordia, n (rate x duration) is

$$n = (T-T_{bi})\hat{\Theta}_s/\hat{\Theta}_i(T-T_{bs}) \qquad (5.7)$$

Therefore, provided the base temperatures for rate and duration are similar, the number of primordia depends only on the ratio of the thermal durations, i.e. $\hat{\Theta}_s/\hat{\Theta}_i$, and is independent of temperature. Important processes governed in this way include spikelet production in some cereals, and flower production in determinate legumes[4].

The actual sink, particularly in reproductive structures, is very much less than the potential near the upper and lower limits of the temperature range for a species. Even a short exposure (for a few hours during one or two nights) to high or low temperatures, at some critical stage such as pollen

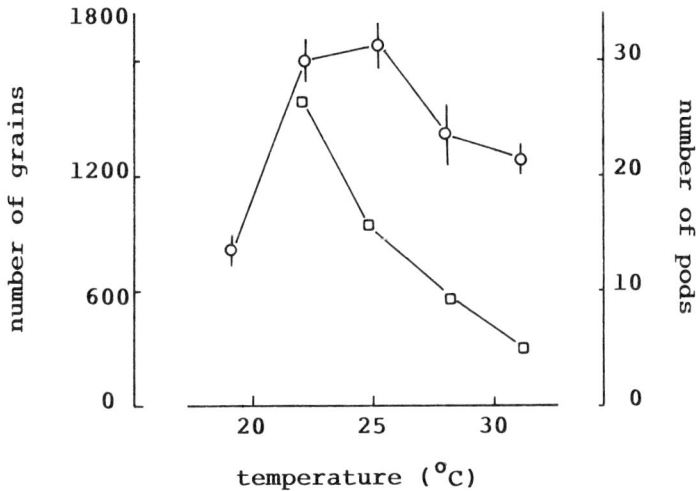

Fig. 5.5. **Temperature and survival of fruiting units.** The number of grains on the main culm of pearl millet (○) and the number of pods on groundnut (□); plants in field soil in controlled environment glasshouses. Sources: Ong, 1983b, 1984.

development, can greatly reduce the number of reproductive primordia[4,5]. There is usually however, a broad intermediate range of temperature in which many primordia survive. Within this range, the number surviving is sometimes moderately but systematically affected by temperature. Most commonly, the number decreases as temperature rises in this range (e.g. the millet in Figure 5.5). The control of this response is considered later.

The effect of developmental duration

The effect on the duration of growth is generally the one dominating partition of determinate structures over the range of temperature where the sink is stable. By substituting the term for t_w (from equation 5.4) in equation 5.1, it is shown that the dry mass of a structure, $W_s(max)$, responds to temperature as in

$$W_s(max) = p\Gamma\hat{\Theta}_w/(T - T_{bw}) \tag{5.8}$$

which implies the dry mass should be inversely proportional to temperature above a base. The response is shown in Figure 5.6a for two stands grown in controlled environment glasshouses. Temperature had little effect on the rate of dry matter production by the stands, so the stand at the lower temperature, taking longer to develop, produced more dry matter, both before flowering (such that W_o was larger) and during panicle filling. The ratio of the durations of vegetative and reproductive phases was little affected by temperature, as was the partition factor, which was close to 1 for panicle growth. Accordingly, the harvest index for panicles was conserved (at about 0.45 – less than 0.5 because there was less dry matter production after than before the start of panicle filling).

In this example, the value of p for panicles was close to 1 because there was a large enough sink, in the form of grains, to receive all the dry matter produced after flowering. In treatments in the same experiment where the minimum night temperature fell to 14°C, this sink was halved (Fig. 5.5), and p for panicles was reduced accordingly. The value of p changed little over the range of mean temperature from 22°C to at least 31°C (the limit of the experiment); but p might be reduced at temperatures higher than 31°C. Genotypes differ much in the width of the temperature range in which p is conserved, as illustrated by the results of altitude trials with maize and rice referred to in earlier chapters.

Examples from altitude trials

The yield of the maize, grown at three altitudes in western Kenya[7], decreased as mean temperature increased from about 15–22°C (the form of response in Figure 5.6a). In this range of temperature, the sink was not strongly suppressed – indicating a high degree of tolerance to low

(a)

(b)

cumulative above-ground dry matter (g m^{-2})

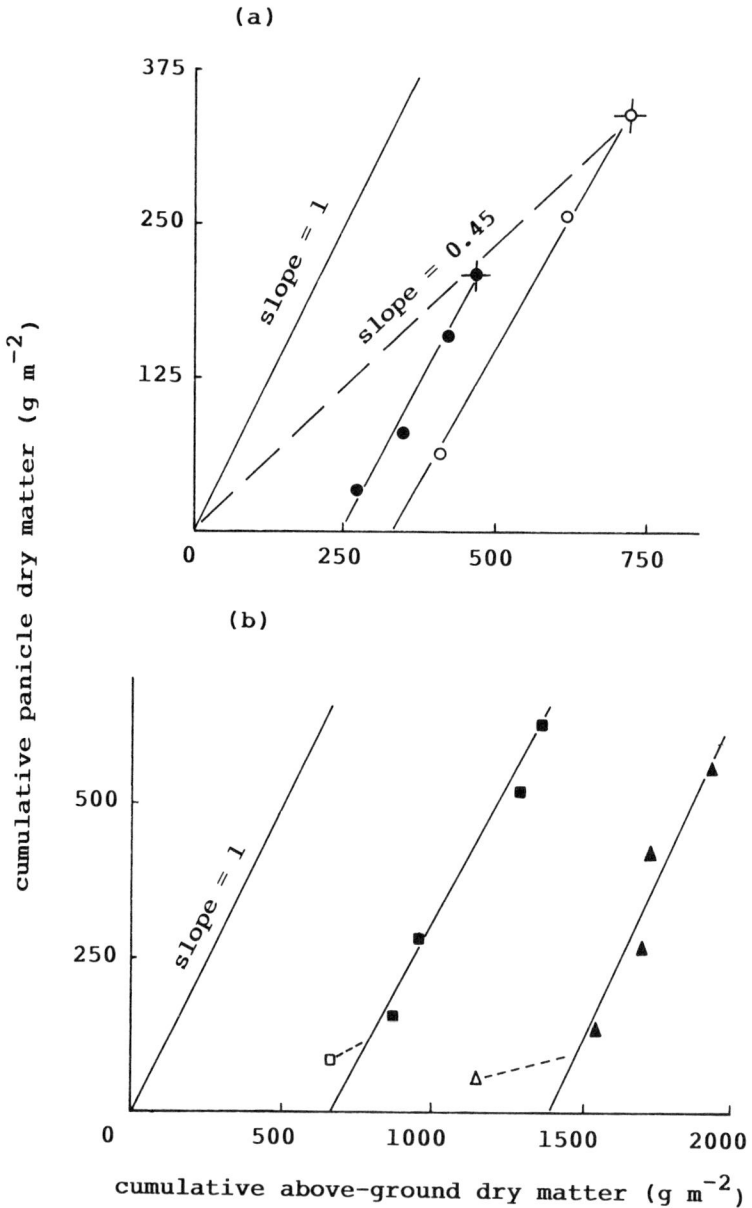

Fig. 5.6. Effects of temperature and photoperiod on allocation of dry matter to panicles in pearl millet. At (a) temperatures of 31°C (●, $W_o = 250$ g m^2, p = 0.91, $r^2 = 0.97$) and 22°C (○, $W_o = 330$ g m^{-2}, p = 0.86, $r^2 = 0.99$) in controlled environment glasshouses, main culm only, ±standard error at harvest, adapted from Squire, 1989b; and (b) photoperiods of 13.5 h (■, $W_o = 680$ g m^{-2}, p = 0.9, $r^2 = 0.98$) and 15.5 h (▲, $W_o = 1410$ g m^{-2}, p = 1.1, $r^2 = 0.93$) in the field in India, for all culms, regressions through closed symbols, original data of Carberry and Campbell, 1985.

temperature by the cultivar used (whose T_b was about 9°C). The number of grains decreased moderately with rise in temperature, but number probably did not govern yield, since the grains at the highest temperature seemed incompletely filled. Most of this response was caused by effects of temperature on the duration of panicle filling (Table 5.2).

Table 5.2. Change with altitude of grain yield and associated attributes for maize in western Kenya.

Site	Chemilil	Kitale	Elgon
Altitude (m)	1268.0	1890.0	2255.0
Temperature (°C)	21.5	17.4	16.0
Duration of grain growth (d)	46.0	53.0	65.0
Grain number per plant	400.0	480.0	590.0
Mean grain weight (g)	0.33	0.35	0.40
Grain yield per plant (g)	132.0	168.0	236.0

Source: Cooper and Law, 1978.

The yield of rice, at three altitudes in Thailand[7], increased as mean temperature rose from 18.5°C (mean minimum, 16°C) to 23.5°C – a response opposite to that of the maize just considered. Despite a potentially longer duration of filling at the lower two temperatures, yield there was restricted because the low temperature caused sterility of florets (Table 5.3). In this example, possibly only the stand at the highest temperature was within the range of temperature where the sink and p are conserved.

Table 5.3. Change with altitude of grain yield and associated attributes for rice (cv. CMU 3) in Thailand, ± standard errors.

Site	Phabujom	Chang Khian	Pa Kia
Altitude (m)	800.0	1200.0	1450.0
Temperature (°C)	23.5	20.5	18.5
Grain yield (kg ha^{-1})	5040±110	1200±510	0
Panicles m^{-2}	192±9	210±10	212±9
Spikelets per panicle	87±3	96±6	110±5
% florets sterile	6±1	80±7	100.0

Source: Chamberlin and Songchao Insomphun, 1982.

Temperature: indeterminate

In indeterminate genotypes, vegetative structures usually continue to grow when fruiting has begun; and new sinks of both types of structure are produced throughout reproductive growth. Therefore, t_w for vegetative processes is independent of temperature (unless this drops low enough to stop development), but t_w for fruiting increases as temperature rises. The ratio of the reproductive duration to the vegetative (i.e. the total) duration also increases with rise in temperature. Accordingly, the potential dry mass of reproductive structures increases with rise in temperature.

The potential number of reproductive sinks similarly increases with rise in temperature between T_b and T_o, as described for leaves and stem internodes in Chapter 2 (equation 2.6). This response governs n in herbaceous annuals, where t_s for the production of leaves, flowers and pods depends largely on the length of the growing season, and also perennial tree crops such as the oil and coconut palms, which continue producing primordia until they are felled. The control of n in branching plants is more complex if some parts of the plant are determinate and others indeterminate. For example, the production of leaves, branches and floral primordia on a main stem of a legume, whose development is indeterminate would be governed by equation 2.6, whereas the production of leaves and flowers on individual branches might be more or less determinate, and governed by equation 5.7. Moreover, in most indeterminates, the time from sowing to first flower usually responds as a determinate period, so the ratio of n for reproductive to n for vegetative structures increases with rise in temperature cf. the indeterminate response for canopy duration and seasonal solar energy interception (Chapters 1 and 3).

In reality, however, other effects of temperature tend to dominate those of duration and the potential number of sinks. These effects on partition result from temperature either directly inhibiting reproductive processes, (e.g. groundnut in Figure 5.5) or promoting vegetative growth at the expense of reproductive, and are now illustrated by two examples, again from the altitude trials described in previous chapters.

Examples from altitude trials

In the first trial, in Zimbabwe, groundnut was grown at 17, 20 and 23°C. The potential reproductive sink increased with rise in temperature – as expected from equation 2.6 – from 700 pegs per m^{-2} field at 17°C to 1800 at 23°C. Only about half the pegs at the highest temperature developed into pods, with the result that the actual sink – the number of kernels – was largest at the intermediate temperature. Yield was also largest there despite the much greater total dry matter production at the warmest site. There are several possible causes of the reduced sink at higher temperature: the pegs, produced on taller stems, had farther to travel to reach the ground; and flowering itself might have been directly impaired by the temperature. In this example, yield was limited at the coolest site because the potential sinks were produced and filled too slowly; and at the warmest site because the number of sinks was reduced though assimilate was plentiful. The partition factor was not conservative, and might rarely be conservative in this species, since work in controlled environments indicates the actual sink would continue to decline further with rise in temperature (Fig. 5.5).

In the second trial in Columbia, cassava grew at 20, 24 and 28°C. As temperature rose, the canopy expanded faster and intercepted more solar radiation over the season, but also required an increasingly greater proportion of the dry matter. For a genotype that was not vegetatively vigorous, the yield of tubers increased with rise of temperature from 20–28°C – the extra dry matter from faster canopy expansion at higher temperature was much more than the increased requirement by the shoots. For another, very vigorous cultivar, tuber yield was greatest at 20°C (where it was three times that of the less vigorous), but decreased to very small values at higher temperatures. For this vigorous genotype, the extra dry matter produced at higher temperature was very much less than the increased requirement by the shoots. (In fact, at the highest temperature, the new dry matter was little more than required to sustain shoot growth.)

Temperature and partition: summary

These examples from altitude trials illustrate the great and diverse effects of temperature on partition of assimilate. The maize, rice, groundnut and the two cassava cultivars all had a different optimum for partition, caused variously by effects on the duration of filling, the actual sink, and the priority other structures had for the assimilate. Effects on duration largely determined the responses of the maize and the vegetatively less vigorous cassava (though in opposite directions). The priority afforded to shoots determined that of the vigorous cassava. In the maize and both cassava cultivars, yield was limited by the supply of assimilate: at the optimum temperature for economic yield, the supply was either available for the longest period or allocated least to other structures. In contrast, a restriction of the reproductive sink dominated the responses of the rice and the groundnut. At higher temperatures, the rice would probably have become supply–limited (as the maize), but the groundnut would probably have remained sink–limited.

In rice and groundnut, certainly, and probably in maize, the transition between supply– and sink–limitation happens over a very narrow interval of temperature (perhaps 2–3 degrees).

Photoperiod

Photoperiod influences partition of assimilate mainly through its effect on the duration of the vegetative relative to the reproductive process. As considered in Chapter 1, photoperiods longer than the critical (for short-day plants, the reverse for long-day) increase the length of the vegetative phase. In determinate cereals[8], long photoperiods have little effect on the length of the reproductive phase, but reduce the ratio of reproductive to vegetative durations; they also have little effect on the partition factor for reproductive growth. Accordingly, long photoperiods increase W_o for reproductive structures and hardly affect the grain yield, but reduce the harvest index (Fig. 5.6b). In contrast, in indeterminate types, the increased vegetative period might cause a shorter reproductive period, and so less reproductive yield, if the length of the growing season is limited.

A secondary effect of photoperiod is to change the partition factors for different vegetative structures. In tillering cereals, the increased expansion and growth of the main culm in long photoperiods reduces the partition factor for the tillers[8]. A comparable effect is particularly strong in cassava (a long-day plant). For example, in Columbia where the days are short (about 12.5 h), artificially increasing the daylength to 16 h reduced the time-to-flowering, and therefore to branching (since branching of the stem occurs when the apex becomes reproductive), and increased the rate of branching. Consequently, there were more apices above-ground under long-days than short, and more of the total assimilate was allocated to these apices than to the storage roots. With reference to the scheme in Figure 5.1b, long photoperiods had little effect on W_o for tubers but reduced the partition factor. The magnitude of this effect differed between clones, and was much greater for those having more vigorous shoot growth.

Drought and partition

Drought has diverse effects on partition of assimilate, depending on its severity and its incidence in relation to the stage of development, and on whether it affects a sink for assimilate or the time during which a sink fills.

Effects on the sink

Drought restricts the potential size and number of units in a structure by preventing initiation of primordia or reducing the rate or duration of initiation. Genotypes differ much in sensitivity, depending on the characteristics of earliness, synchrony of reproduction throughout the plant, and the ability to suspend reproduction then continue it when a drought is relieved[9]. For example, some early pearl millets are relatively insensitive to drought, whereas certain sorghums produce negligible grain if the drought is severe at flowering, even if the plants have produced much dry matter before flowering.

Many indeterminate legumes display characteristics that combine both sensitivity to drought and an ability to survive drought. The sensitivity is commonly the result of reproductive processes being synchronized with leaf production, which itself is very sensitive. Consequently, drought both increases the thermal time to the start of flower initiation (and thereby shortens the overall duration of initiation) and slows the rate of initiation. For example, at Hyderabad in India, the thermal duration for initiation of a groundnut flower increased from about 10°Cd to 30°Cd as leaf water potential fell from about -0.6 MPa to approximately -1.5 MPa. At leaf potentials lower than about -1.7 MPa, leafing and flowering ceased (Chapter 1, Reference 18). Drought also slowed the rate at which pegs elongated, and it hardened the soil surface thereby preventing them penetrating it. Some of these same genotypes also had the ability to continue producing primordia following the onset of rain or irrigation, even if they had not produced any for several weeks or months; the thermal duration was much the same as before the drought began.

A genotype whose reproductive development is spread over a long time and delayed by drought, is more suited to an environment with a bimodal rainfall distribution, or if the season's rainfall is sporadic, interspersed with dry periods. The same genotype might fail to produce a reproductive sink if there was no rain after the stands were established. On the other hand, a genotype whose reproductive development is compressed and little delayed or impaired by drought, will be more successful growing on early, concentrated rainfall than sporadic or bimodal rainfall.

The physiological linkage between drought and sink size is very complex, and presents problems comparable to those for leaf expansion. There are probably both direct effects of reduced turgor potential on the initiation and survival of primordia, and indirect effects operating through the rate of dry matter production or hormonal signals from other parts of the plant.

Effect on the duration of growth

It was described in Chapter 1 that drought can delay the time to the point at which structures begin to fill, and cause structures to cease filling prematurely. Both effects tend to reduce t_w for reproductive growth as a fraction of total growth. Again, species, and even different structures on a plant, differ much in their sensitivity. Short-season dryland crops, such as pearl millet and cowpea, are among the least sensitive. Provided the root system can extract water until at least the plants set grain, the start of reproductive growth is not delayed (and might even be accelerated). Moreover, the duration of reproductive growth might still be similar to the value for watered stands, even when there is no net dry matter production after flowering. In contrast, reproductive development in other cereals and legumes might be suspended indefinitely during drought. For these genotypes, the ability of the foliage to survive drought will determine whether reproductive growth proceeds if the drought is relieved.

Partition between roots and shoots

Root systems have not been considered so far in this chapter, largely because there are few relevant measurements. The fine root system is a small proportion of the total dry mass (commonly 5–15%) when drought is not limiting growth, and so will have little bearing on the responses to temperature and photoperiod. Roots constitute a much larger proportion of the total dry matter in dry environments.

In a plot of root against total dry mass, W_o is zero and p for roots is usually fairly stable initially while the canopy and root system are expanding (cf. Chapter 2). In determinate plants, p for roots usually decreases when the stems begin to grow rapidly. After flowering, root systems cease extending while shoots remain a sink for dry matter. The root to total dry weight ratio for cereals is therefore largest during the early period of expansion, and decreases with time. For example, the root to total dry mass ratio of barley in one experiment in Syria decreased threefold from the beginning of stem extension to anthesis[10]. In indeterminates, both root and shoot systems can accrue dry matter throughout development, but p is not always constant (groundnut[10]).

It was shown in Chapter 2 that, when plants take their water from a drying soil, the length of the root system increases, while the area of the canopy decreases. These responses sometimes cause a slightly greater root mass in a drying soil, but usually a much greater root to total dry mass ratio. In the series of experiments with a short-season pearl millet referred to in earlier chapters, p for roots changed from 0.1 for an irrigated stand at Hyderabad to 0.34 for a stand growing on stored water on a deep sand in Niger. The change in p more or less compensated the change in total dry mass, such that the growth rate of the root system between sowing and anthesis 45 d later, changed little from 0.8 to 1.1 g m^{-2} d^{-1}. The corresponding duration of growth was unaffected by drought, so the final weight of roots was similar at both sites, though root weight as a fraction of shoot weight increased with increasing dryness. Similarly the ratio for groundnut in the dry season at Hyderabad was more than twice that in the monsoon (Chapter 4, Reference 6).

Some dryland crops achieve an even larger root to total dry mass ratio than the millet in the above example. The ratio before anthesis for barley in Syria[10] varied between 0.3–0.49 in different treatments and localities, and was usually larger in drier, as opposed to more moist, environments. Such responses are not restricted to the fine root system: the partition factor (but not W_o) for cassava tubers increased during drought in

Columbia[10]. It increased from 0.4 to 0.71 in a vegetatively vigorous cultivar, and from 0.76 to 0.89 for a much less vigorous one. This effect is similar to the effect of low temperature described earlier. Both low temperature and drought increased p for tubers by reducing the expansion rate (and thereby the sink) of the shoot system.

Several physiological mechanisms could cause the increase of p for the fine root system in dry environments. Provided a root system is extending into moist soil, its growing roots will probably be more turgid than leaves directly exposed to the dry atmosphere. More generally, hormonal interactions between root and shoot are inferred.

Partition between vegetative and reproductive structures

Partition between these structures depends much on the timing of the drought, and on certain physiological attributes of the genotype. There are many types of effect on W_o and p, resulting from combinations of environment and genotype.

Two of the most common effects were observed in an irrigation experiment using sorghum in Nebraska, USA (Fig. 5.7). In one set of treatments, stands were grown with different amounts of irrigation, or with

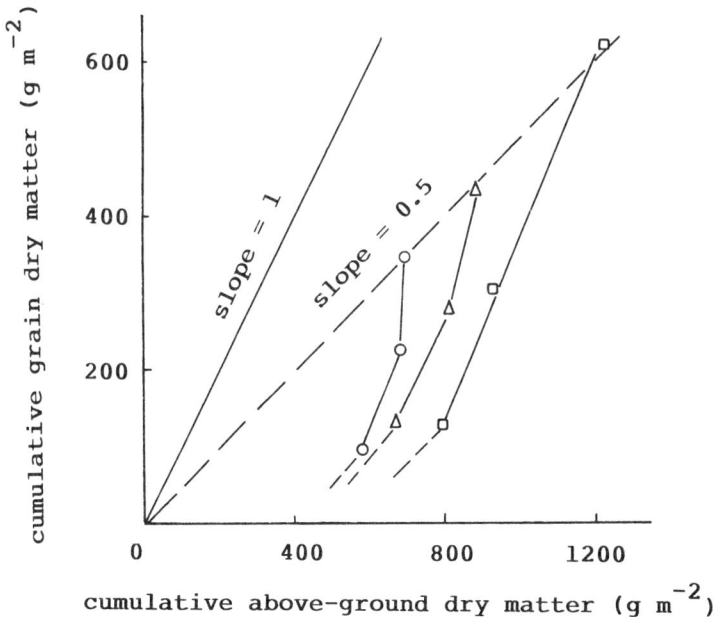

Fig. 5.7. Effects of drought on allocation of dry matter to grain in sorghum. In Nebraska, USA, not irrigated (○), or irrigated to replace 80% (□) or 40% (△) of the water estimated lost by evaporation from all surfaces. Data of Garrity, Sullivan and Watts, 1983.

no irrigation. There was enough water in the soil, even when there was no irrigation, to enable plants to set a reproductive sink. Stands receiving little or no water grew more slowly before flowering than those that were fully irrigated. Their rate of dry matter production was reduced further during reproductive growth, but the partition factor for panicles increased to much greater than unity – implying retranslocation of dry matter from storage organs to the panicle. The harvest index was conserved, despite a twofold difference in total dry matter between stands. Another set of treatments (not shown) was fully irrigated before flowering, and droughted to different degrees thereafter. For these treatments, W_o was similar, but those receiving less water produced less dry matter during panicle filling. The more droughted stands retranslocated assimilate, thereby increasing p and again conserving the harvest index.

The increase of p in response to drought in these examples is a widespread effect in the dry tropics. The ability to retranslocate dry matter stored in the stems is an important attribute, but one in which genotypes differ greatly[11], though little is known of how they differ at the cellular level. Certain millets and sorghums can provide up to 100% of the dry matter for panicles from dry matter stored in the stems; others seem able to retranslocate little more than 20%. Some legumes might move small amounts, but most seem unable to move any. Perhaps more commonly, p is reduced as a result of drought, because either only a small sink is set, or the genotype is unable to use stored assimilate to fill the sink.

Nutrients

Few of the effects of nutrients on partition of dry matter are understood at the level of the stand. The following summarizes some of the most common effects on the sink, on the duration of filling and on partition factors.

Effects on the sink

If nutrients are added to a soil, the number of sinks for dry matter in vegetative and reproductive structures usually increases. Effects on the vegetative sink result in more axillary structures, such as tillers or branches, or at least influence the number of leaves and internodes on axillary structures more than on the main axis. In deep-water rice, this effect is sometimes pronounced because the longer stems of fertilized plants reduce the chance of the culms being completely submerged and rotting if the water level rises[12].

The duration for which primordia are initiated is much less affected than the corresponding rate of initiation, or the fraction of initiated primordia that survive. The suppression of the number of reproductive units by nutrient shortage is widespread. It is difficult, however, to determine whether nitrogen shortage affects the sink directly, by impairment of the metabolic processes in the differentiating units, or indirectly, through the reduction of dry matter production in the whole plant.

Effects of nutrients on partition

Shortage of nutrients increases the proportion of dry matter allocated to roots compared with aerial structures, an effect consistent with that on the R/L ratio described in Chapter 2. The effect can be large: for the barley in Table 2.2, the root/total dry weight ratio during canopy expansion decreased from about 0.5 in unfertilized soil to 0.29 at one site with added nitrogen and phosphorus[10]. Generally, however, the increase of p for roots in nutrient deficient soil does not compensate for the much larger concomitant decrease of total dry matter production, and a genotype produces less dry matter in roots in nutrient-poor than in nutrient-rich soils.

Studies of the effects of nutrients on partition in the shoots seldom take account of these large changes below ground. Above ground, a shortage of nutrients has a range of effects on partition between vegetative and reproductive structures. A deficiency sometimes decreases the fraction of dry matter allocated to reproductive structures and occasionally increases this fraction but mostly has little effect on it. The existence, and direction, of an effect depends on genotypic traits and on other environmental factors, as shown in the following examples.

The first example shows the effects of a set of fertilizer treatments on the allocation of dry matter to grain in maize (Fig. 5.8). Increasing the dose of nitrogen from 0 to 42 g m^{-2} more than doubled W_o, increased grain yield about fourfold and the harvest index 1.5 times, but hardly affected the partition factor for grain. Since p was conserved, nitrogen must have influenced partition through its effects on development or the rate of dry matter production, or both (Chapters 1 and 3). Generally, effects on the rate of production were greater than on development. Even though the vegetative phase became shorter as more N was applied, the concomitant increase in the rate of dry matter production was enough to cause the change of W_o. The larger effect of N on grain yield was the result of both a longer period of filling and more rapid dry matter production as the dose of N was increased.

The partition factor was slightly smaller in the treatment with no N than in the other treatments, but was about 1 in all treatments. (The cobs were the main sink for dry matter during reproductive development.) The number of grains per cob increased as the dose of N was increased, and the size of the sink seemed not to limit partition in any of the treatments.

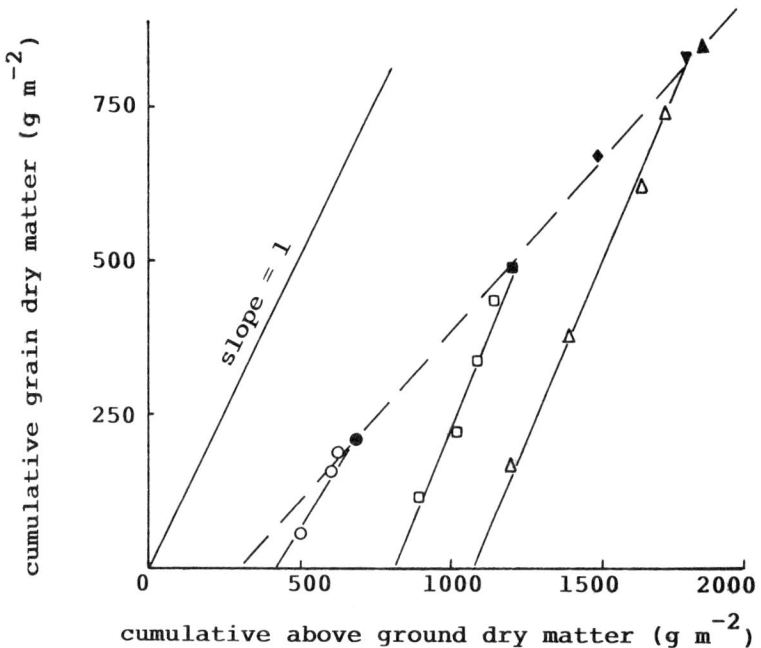

Fig. 5.8. Effect of nitrogen fertilizer on the allocation of dry matter to grain in maize. On a red clay loam, in Australia, with no added fertilizer ($\circ$, $\bullet$) and with 6 ($\square$, $\blacksquare$), 12 ($\blacklozenge$), 24 ($\blacktriangledown$) and 42 ($\triangle$, $\blacktriangle$) g N m^{-2}. The closed symbols are values at maximum dry mass (at harvest or just before); the open symbols, shown for three treatments only (for clarity), are values at sequential samples after anthesis. From the linear regressions: for the closed symbols, intercept is 350 g m^{-2} (plant population 7 m^{-2}), slope is 0.57; for the open symbols, p is 0.8, 1.2 and 1.2 for 0, 6 and 42 g N m^{-2}, respectively. Original data of Muchow, 1988b.

Grain number was well-matched with the available dry matter, such that the dose of N had little effect on the mean rate at which grains filled. The mean mass per grain also increased, from 200 to 270 mg, as the dose of N increased; a response entirely due to the lengthening of the filling period. This effect on duration largely caused the comparable effect on harvest index. It is uncertain how widespread this effect is in maize, but it has been observed elsewhere[15].

The information available suggests partition is more sensitive to fertilizer in maize than in other cereals. For sorghum in the same experiment as the maize in Figure 5.8, the dose of fertilizer hardly affected the duration of grain filling and p for grain. Consequently, the mean grain mass and the harvest index changed little. Such stability of the harvest index is also shown in experiments with sorghum[16] and rice (Fig. 5.9a) in India. Dry matter during growth was not measured in either of these

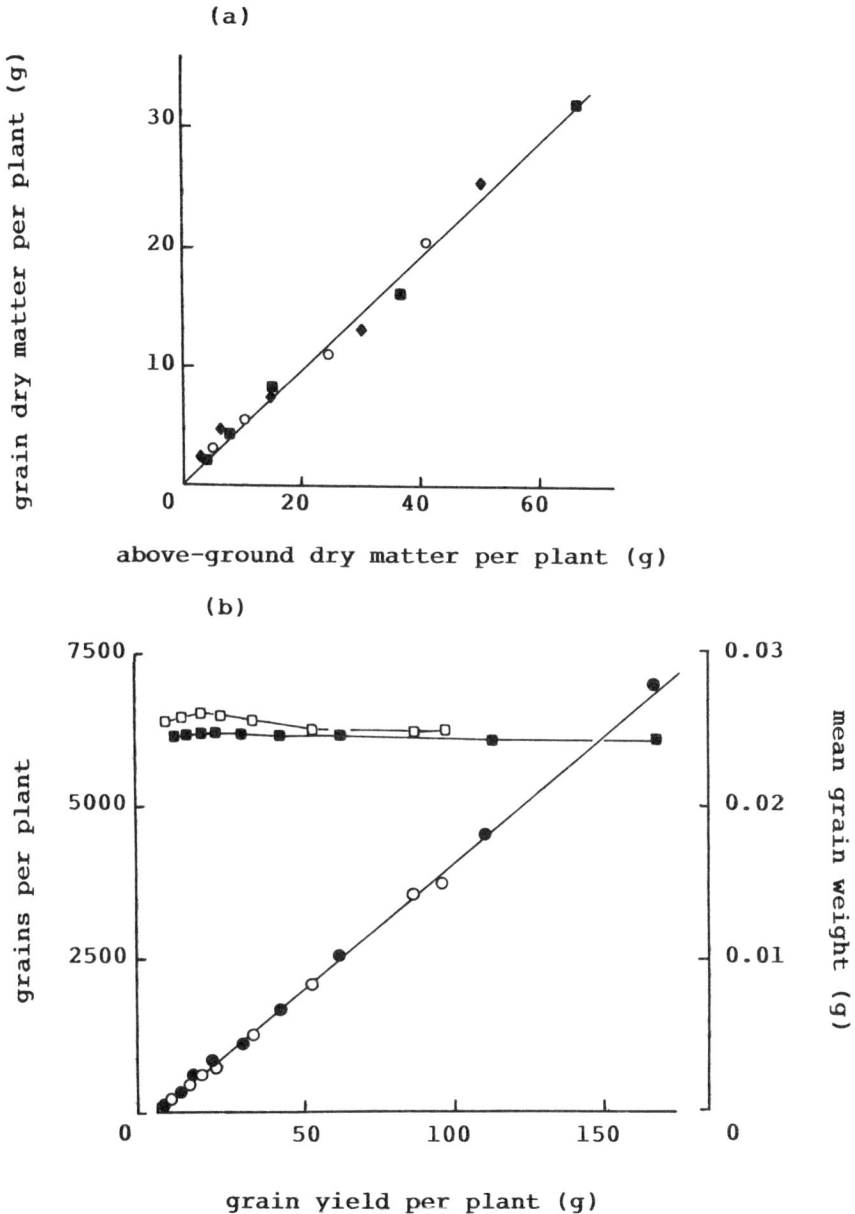

Fig. 5.9. **Effects of nutrients on the response of grain yield and its components to plant population in rice.** Increasing plant population caused the reductions in total dry matter and grain yield per plant. Part (a) shows dry matter at harvest for each population, without nitrogen (○), and with 60 (♦) and 120 (■) kg N ha^{-1}, in Uttar Pradesh, India; slope is 0.48, from Pyare Lal *et al.*, 1982. Part (b) shows effects on grain number (circles) and mean grain weight (squares) of heavy manuring (●, ■) compared with only a basal dressing (○, □), from data of Akita, 1982c; the line is 41 grains g^{-1}.

experiments, but the results imply nutrients had little effect on the duration of panicle-filling and the partition factor.

The effects of fertilizer on partition in legumes are difficult to interpret. The total amount of dry matter they produce is not strongly affected by varying the dose of nitrogen. It is therefore difficult to analyse graphical plots of grain on total dry matter at harvest since most of the data are clumped. There is some evidence that a smaller fraction of the dry matter is allocated to grain as the dose of nitrogen is increased[16]. But this was not the response of a non-nodulating cultivar of groundnut, which is unable to fix atmospheric nitrogen[16]. This cultivar appeared to respond like sorghum or rice, in that increasing the nutrient dose ($0-200$ kg N ha^{-1}) increased dry matter production but had a negligible effect on the harvest index, i.e. a plot of grain on total dry mass at harvest was linear with an intercept very close to zero.

In these examples, the nutrient treatments had only small effects on partition of dry matter between vegetative and reproductive parts of the plant. Circumstances arise, however, when fertilizer increases the fraction of the dry matter allocated to vegetative structures. This is most likely to occur when nutrients are applied very early in vegetative expansion and when another constraint reduces dry matter production during reproductive growth. For example, if a stand is growing on the store of water in a soil, the additional fertilizer during vegetative expansion will increase the rate at which the store is used, with the result that there might be little water left for reproductive growth. This effect of nutrients would be expected to be lessened for cultivars which can set a large reproductive sink in dry conditions, then retranslocate stored assimilate to fill that sink.

A final example shows a different effect of nutrient shortage: one that *increases* the allocation of dry matter to reproductive structures. A wide range of total dry matter was obtained among the plots of a fertilizer trial with oil palm. As in response to increasing density with this species (Fig. 5.4a), nutrient shortage reduced bunch dry matter proportionately more than vegetative dry matter – but the latter was far from conservative as in response to density. Here, nutrient shortage might have increased p for fruits by suppressing the expansion of fronds, which normally have priority for dry matter.

These and other studies show that adding nutrients to a soil stimulates yield by raising dry matter production more than the harvest index. Even an effect on the harvest index could be a response to the change in dry matter per plant (as in Fig. 5.4c). Whether the harvest index is reduced by this response, or by another response to nutrients, can be shown by comparing the relation between grain and total dry mass when nutrient dose is the variable, with the corresponding relation when plant population is the variable and nutrients are not limiting. If the relations are similar, nutrients are affecting the harvest index entirely through mean plant dry mass. If the intercept of the relation (on the x–axis) is larger when nutrient dose is the variable than when population is, nutrient shortage is reducing the partition to reproductive structures independently of the effects on plant dry mass (and vice versa).

For example, the relations for rice in Figure 5.9a were similar whether population or nutrient dose was the variable, indicating that nutrients affected dry matter production only. There are no such side by side comparisons for the maize in Figure 5.8 or the oil palm in Figure 5.10. Other physiological evidence suggests nutrient shortage decreased the harvest index of the maize and increased that of the oil palm. Consistent with this evidence is that, for the maize in Figure 5.8, the intercept of the line (for data at harvest) is, at 50 g per plant, much greater than the value of W_v for maize in population experiments; and for the oil palm in Figure 5.10 the intercept of 12 kg per plant is very much smaller than the 80 kg for W_v in Figure 5.4a.

Nevertheless, these effects on partition in maize and oil palm are still smaller than accompanying effects on production. It seems nutrients generally affect partition less than production because plants can move nutrients between structures[18] more readily than they can move stored carbohydrate. In some species such as pigeon pea[11], the retranslocated nutrients account for all the nutrients in the grain. The movement of nutrients from leaves during grain filling is commonly associated with a

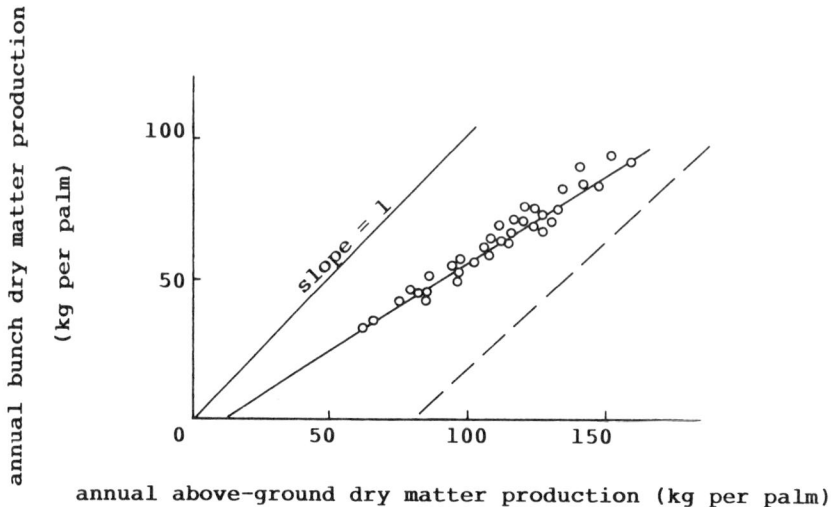

Fig. 5.10. Effect of fertilizer on allocation of dry matter to fruit bunches in oil palm.
Plots in an NPK fertilizer trial, on an infertile soil in peninsular Malaysia. Dry matter corrected for energy content of oil. Population density, 148 ha^{-1}. From the linear regression, the intercept is 12 kg per palm, and slope 0.64. The dashed line is the slope for fruit bunches in Figure 5.4a. From the same experiment as Figure 3.12 (Chapter 3, Reference 25).

decrease in the rate of photosynthesis and this, rather than a shortfall of nutrients as constituents of the grain, generally limits reproductive yield when nutrients are scarce.

The influence of cultural factors

There is some information to show that cultural factors, such as plant population and the composition of the stand can modify the adverse effects on partition of drought and nutrient shortage.

Plant population density

A sparse stand will use the store of water in a soil less rapidly than a dense one (Chapter 4), and so will have a greater chance of setting and filling reproductive structures. The sparse stand will therefore have a greater partition factor for grain and will accumulate relatively more dry matter after grain filling begins. Because of these effects, the harvest index in dry conditions usually increases as population density decreases. (Drought usually increases the value of W_v derived from the relation between grain and total dry matter in a population experiment[19].)

This advantage of reducing population density is, to some extent, offset by the larger fraction of dry matter allocated to roots at low population. For example, the partition factor of 0.36 for groundnut roots given earlier was for a stand at 22 m^{-2}; the ratio increased to 0.5 at a population density of 7 m^{-2}[10], consistent with the corresponding differences in root length and leaf area (Tables 2.2 and 4.3).

There are many reports that population influences the response to nutrients, but few examples to show systematic effects on partition. Only grain yield is measured in most experiments. Data for rice (Fig. 5.9a), for which W_v is generally small, indicate very little effect of population on the relation between grain yield and total dry matter per plant in each of three nutrient treatments.

Mixed cropping

Two general ways of mixing species have been examined. In one, a row of one species is replaced by a row of another; for each species in the mixed crop, the distance between plants in a row is the same as in the respective crop grown alone, and the population density is similar in all crops. In the other, the species are combined at their densities when grown alone. (One of the stands is usually at a wide row spacing to allow insertion of the other.) The population in the mixed stand is therefore much greater than in either of the others.

Most of the cereal/legume mixed crops examined in earlier chapters were of the first type (little change in population), and most were grown in the moist conditions of the Indian monsoon (Chapter 3, Reference 16). For the legume, mixed cropping had little effect on the timing of processes, the dry matter produced by a plant or the partition factors of the main structures, and so the harvest index was unaffected. For the cereal, mixed cropping increased the dry matter production per plant. If the cereal is one that tillers, much of the additional dry matter was allocated to tillers; hence p for the main culm decreased, though partition between reproductive and vegetative structures in a culm was little affected.

In moist conditions, partition seems also little influenced by the other form of mixed cropping. In a trial with sorghum/cowpea mixtures in Australia[20], both the population density of the cowpea and nitrogen fertilizer were variables (the density of the sorghum being constant). The combinations ranged from no fertilizer and 8 m^{-2} cowpea, to 120 kg N ha^{-1} and 15 m^{-2} cowpea. None of the treatments had much effect on the harvest index of the sorghum; a plot of grain on total mass at harvest is more or less linear, with an intercept near zero. The interpretation of the responses of the cowpea is uncertain, since the nitrogen treatments had little effect on its total production.

The conclusion from these studies is that intercropping in moist environments has little effect on the fraction of total dry matter allocated to grain. The species, sorghum, pearl millet, pigeon pea, groundnut and cowpea have a stable partition factor which enables them to conserve the harvest index in the face of large changes in total plant mass.

In dry conditions, there is some evidence that mixed cropping can have large effects on partition between grains and the rest of the plant. In a range of dry treatments[21], the harvest index of sole crops was reduced during drought for both sorghum and groundnut, though not for millet (a difference consistent with many of the attributes of these species). The harvest index was less affected when either the sorghum or groundnut were mixed with one of the other two species. The physiological mechanisms influenced by intercropping were not discovered.

Relations between number, size and mass

A change in the dry mass of a structure, caused by any of the factors considered in the previous section, is accompanied by changes in the size and number of the units that constitute the structure. Size and number are important for reasons other than that they might themselves determine the amount of dry matter allocated to the structure. Each reproductive unit has a certain amount of surrounding vegetative material, in which dry matter is expended irrespective of how much is allocated to the edible part inside. Accordingly, a few well-filled units are a more efficient use of assimilate than many partly-filled ones. Moreover, the size and mean mass of units has practical significance in the kitchen and market place. This section now considers systematic effects on number, size and mass.

Compensation in number and conservation of unit mass

When environmental or cultural factors restrict the amount of dry matter accumulated by stands or individuals, physiological mechanisms respond to conserve the mean mass of the sub-units in their main structures. Compensation is usually only partial in vegetative parts: both number and mean mass decrease. Compensation is often nearly complete in reproductive parts: number decreases while mean mass is conserved.

This conservatism is best shown when rising population density is the factor forcing down the mean plant mass, as in Table 5.4[22]. The mass of grains, pods and fruits is less stable if plant mass is restricted by temperature, drought or shortage of nutrients.

Table 5.4. Effects of population density on attributes of seed yield in pigeon pea.

Population (m^{-2})	Dry matter		Seed number per plant	Mass per 100 seeds (g)	Seed nitrogen (%)
	total	seed			
	(g per plant)				
0.67	2090	380	4360	8.76	3.55
1.2	1500	220	2460	9.06	3.55
2.69	780	84	990	9.52	3.72
5.38	430	31	310	9.79	3.65
21.5	110	7.1	70	10.26	3.80

Source: Akinola and Whiteman, 1974.

Compensation in response to temperature

Low or high temperature, near the limits of the range in which a genotype will develop, usually suppresses the number of reproductive units relatively more than the rate of dry matter production. Near these limits, the source of assimilate is large compared with the sink, and the reproductive units might reach their maximum mass. This was so for the millet grains at the lowest temperature in Figure 5.5, and the groundnut at the highest.

At moderate temperatures between the extremes, many reproductive units survive, but the number that do is systematically affected by the temperature. The effects are best understood for the grains of some of the main cereals. It was shown previously in this chapter that the potential number of spikelets is stable at these moderate temperatures, whereas the amount of dry matter in the reproductive parts of the plant decreases as temperature rises to T_o. (This effect on dry matter is governed mainly by the response to temperature of the duration of reproductive growth.) The number of grains that are set and grow also decreases as temperature rises. This trend is seen in the altitude trial in Table 5.2. Consequently, the mean grain mass is much more stable than the grain number or the dry mass of cobs or panicles. In several studies, mean grain mass also decreases slightly or moderately with rise in temperature, though the physiological basis of this response is not clear[23].

Compensation in response to drought and nutrient shortage

Compensation by reduction of number to achieve stability of mean mass is still successful in many dry, infertile environments that greatly reduce dry matter production. Some genotypes, such as cowpea, achieve a good match between number and mass in most dry environment[24]. One of the characteristics of these types is that their reproductive sink develops gradually, and each unit of this sink is filled soon after it is set. In the example in Figure 5.11b, mean grain mass differed between the treatments, but was still largely conserved over a fiftyfold range of variation between stands in the total dry mass of grain.

The cereals are less able to maintain stability. All the units of the reproductive sink on a panicle are set within a few days, and require assimilate more or less simultaneously over a period of several weeks. Grain mass is least stable when a stand growing on a reserve of water in the soil exhausts the reserve just after anthesis: relatively many grains are set compared with the assimilate available during grain filling, and mean grain mass is reduced. This happened in the time-of-sowing trial with sorghum in Nigeria, referred to in Chapters 1 and 2. The later sowings filled grains when soil water was in short supply, and total dry matter production decreasing. Mean grain mass decreased as more of the period of grain filling occurred during the end-of-season drought; the relation between grain yield and grain number was not linear as for cowpea (Fig. 5.11a). Whatever the response to drought, grain mass tends to be conserved more in those genotypes able to retranslocate stored dry matter.

There is also stability of unit mass in relation to change in the amount of nutrients in a soil. Compensation is more effective for reproductive structures – the vegetative tend to decrease in mass when there are fewer nutrients. Figure 5.9b is an example of very great stability in rice. Effects of two nutrient treatments were examined over a range of plant popu-

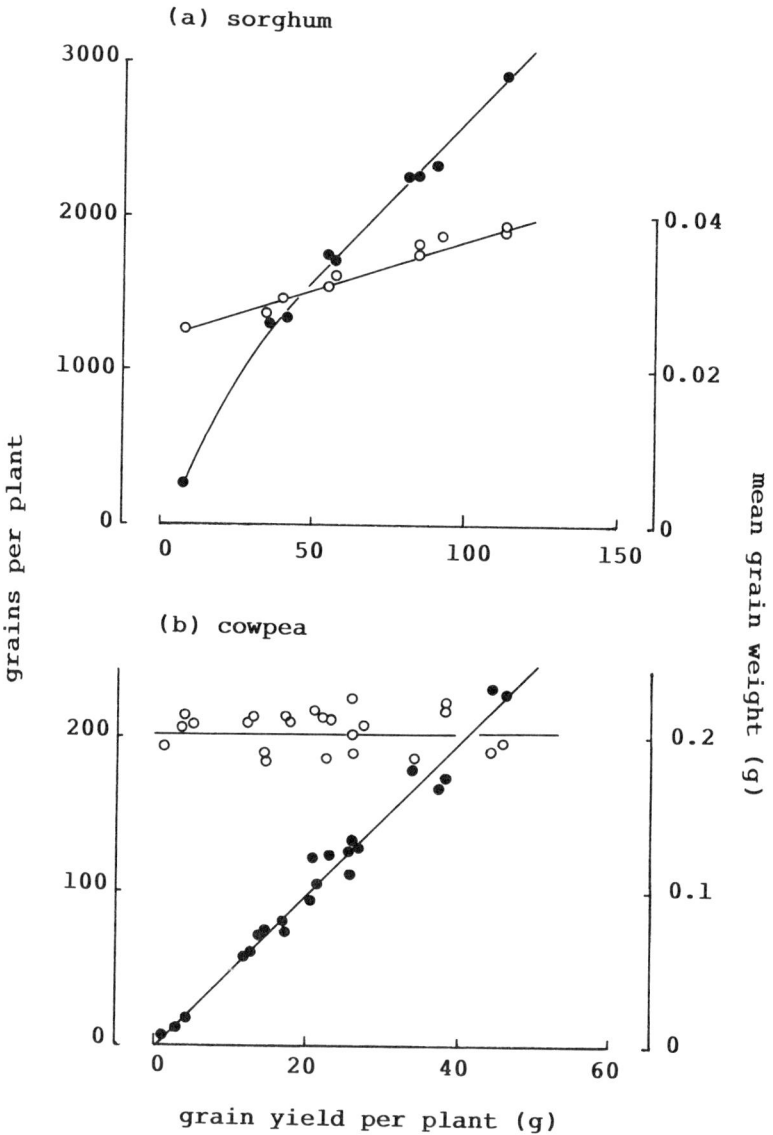

Fig. 5.11. **Different effects of drought on the components of seed yield in sorghum and cowpea.** Graphs show grain yield in differently droughted treatments in relation to grain number (●) and mean grain weight (○): (a) sorghum at Samaru, Nigeria, from data of Kassam and Andrews, 1975; (b) cowpea in California, USA, from data of Turk, Hall and Asbell, 1980. Text gives discussion.

lation, but neither nutrients nor population had much effect on the mean grain mass.

In most experiments in which mean plant mass varies, the number of surviving reproductive units tends to be well-matched with nutrient uptake by the plant, such that the nitrogen content per reproductive unit is conserved (e.g. Table 5.4). Even when nutrients are the experimental variable, the nutrient content of fruiting structures changes much less than many other attributes. The percentage N content is commonly stable over a wide range of applied nitrogen, but sometimes increases when very large amounts are supplied (for example, References 12, 16, 17). The ability of many plants to mobilize stored nutrients more than stored carbohydrate is probably the basis of this conservatism in the fruits (and the variability of nutrient content in leaves and stems).

Matching sink with source: response to thermal growth rate

The direct effects of temperature and drought on survival of primordia account for the imbalances between sources and sinks for dry matter near the environmental limits of the range for a species. Between these limits there is a wide 'window' of conditions in which temperature, drought and nutrients do not have devastating effects on the survival of units, yet within which, survival often varies systematically with environmental conditions. Within this window, the sink for dry matter is usually well-matched with the source, and the mean mass of reproductive units is conserved.

Observations of many species in many environments (within the window frame) have shown that more primordia survive if (a) the plants accumulate dry matter more rapidly, or (b) they develop more slowly[25]. So, for example, the number of units is reduced if plants are shaded, which reduces the rate of dry matter production per plant without affecting the period over which a unit develops. Number is also reduced if determinate plants are grown at a higher temperature (between T_b and T_o), which reduces the period of development, while hardly affecting the rate of dry matter production. The responses are not confined to reproductive processes: fewer tillers per plant survive and grow if plants are shaded or develop rapidly at high temperature.

These responses to rate of dry matter production and rate of development were consistent and systematic in a series of experiments with maize in western Kenya[25]. Among treatments that included weeding, plant population density, time of sowing and altitude, grain number differed over an eightfold range (from 100–800 per plant), and was related to an index – the *thermal growth rate*[25] – defined as the rate of dry matter produced per unit thermal time above a specified base temperature (Fig. 5.12a). (The units of thermal growth rate are dry matter production per unit ground area or per plant per degree-day.) When the conversion ratio

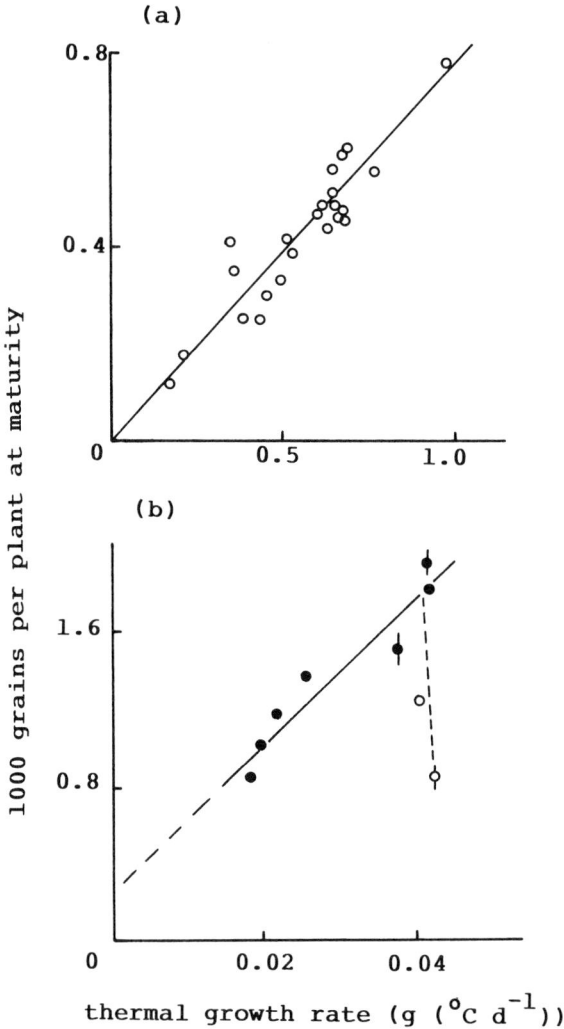

Fig. 5.12. Thermal growth rate and grain number in maize and pearl millet. Thermal growth rate is the dry matter production (per plant) per unit thermal time above the base temperature (T_b): (a) maize during the 'linear' phase of growth, in a series of trials in Kenya (altitude, population, etc., see text), $T_b = 9°C$, $y = 0.73x$, $r^2 = 0.92$, from Hawkins and Cooper, 1981; (b) pearl millet during the period from panicle initiation to anthesis, at Hyderabad and in controlled environment glasshouses, $T_b = 10°C$, for closed symbols $y = 0.27+37.8x$, $r^2 = 0.86$, s.e. of slope is ±6.9, open symbols show low temperature effects on viability (see text), adapted from Ong and Squire, 1984.

is conservative, the index can sometimes be reduced to a 'thermal interception rate', to emphasize that temperature primarily affects the period of time, and thereby the amount of solar energy the plant intercepts while its spikelet primordia differentiate.)

A response of grain number to thermal growth rate was also found for pearl millet, grown in controlled temperature glasshouses, where temperature was the variable, and in the field in India, where growth rate was the variable (Fig. 5.12b). Similar responses to thermal growth rate might explain many other effects of environment on grain number. For example, the relation in rice between spikelet number and nitrogen content of shoots referred to earlier[14], differed with locality in Japan. More spikelets were produced per unit applied nitrogen in cool, northern areas – where the thermal growth rate was probably greater – than in warm southern areas. The work with millet shows how the relation breaks down when extreme temperature suppresses the sink near the limits of the range: in two treatments at mean temperatures of 22°C and 19°C (night minimum 17°C and 14°C, respectively) there were few grains despite the high thermal growth rate.

In these cereals, grain number is proportional to thermal growth rate before flowering; later conditions seem to have little effect on the number. It seems the *number* of surviving grains is determined on the basis of a 'prediction' of the *amount* of assimilate that should be produced during panicle filling. Similar predictive responses probably operate in crops other than the cereals, though many indeterminate species have the advantage of being able to regulate number at several stages of development, thereby compensating for any change in the environment after the first prediction has been made. Oil palm is a species with several regulatory mechanisms[26]. The number of fronds produced in a given time sets the limit for the potential number of fruit bunches (which arise in the frond axils). The first predictive mechanism operates about two years before flowering, when the gender of the inflorescence is determined; poor conditions at that time, reducing dry matter production, lead to more male than female flowers. Several subsequent mechanisms, including abortion of female inflores-cences, bunches and fruits, are increasingly less predictive and more a fine-tuning to immediate conditions.

The predictive mechanisms of different species might differ mainly in the length of time between the setting of the sink and its filling. The longer this time, the more likely the prediction will be wrong, and the sink less well-matched with the source. Among annuals, this time is longer for cereal grains than most legume pods. Accordingly, as the examples in Figure 5.11 show, grain number is less closely related to dry matter produced for cereals than legumes. Errors in prediction can to some extent be compensated for if the genotype can retranslocate dry matter from stems and other structures. Few legumes seem able to do this, so for them, both pod number and the dry matter in pods are commonly determined by conditions after reproductive growth begins, not before.

Control of the partition factor

Many of the examples in this chapter show that the partition factor, p, is a stable quantity both during growth of a stand and in the face of much environmental change. The partition factor does however, respond to adverse environments in all species and to most environmental change in some species. The following briefly summarizes some of the factors controlling it.

For many plants, p is stable when the sinks for dry matter in the different structures are balanced. When environmental constraints reduce the number of surviving vegetative or reproductive units well below the potential number, the small size of the resulting sink itself limits the fraction of the dry matter allocated to the structure. This occurs most commonly in extreme environments, for example when low temperature or drought impairs or prevents grain set. It occurs occasionally in moderate environments, for example when the anthers of wind pollinated plants are drenched by heavy rain. In these conditions, p is usually proportional to the size of the sink, for example the number of growing grains or fruits. The analysis of thermal growth rate is a means of determining when factors such as temperature, drought and nutrients are affecting the sink directly rather than through assimilate supply (e.g. Fig. 5.12b).

More commonly, in moderate environments, the value of p for a structure depends on factors such as determinacy and what is sometimes referred to as 'sink strength'. In determinate types, the growth of the different sinks on a culm is more or less discrete. In genotypes with only one culm per plant, most of the assimilate produced during expansion and growth of each structure is allocated to the structure (p is therefore about 1).

The control of partition in indeterminate types with simultaneous vegetative and reproductive expansion is more complex. In types such as small annual legumes, vegetative and reproductive primordia are initiated synchronously. Environment affects both types of primordia to a similar extent; the two sinks remain balanced, and the partition factors are stable. In other types such as the perennial oil palm and cassava, the above-ground vegetative sink is by far the strongest. If the total dry matter production decreases, the absolute amount of dry matter allocated to this sink might change little, while the partition factor increases. The environment sometimes favours the weaker sink by suppressing the stronger. When this happens, it is *expansion* of the vegetative sink in the shoots that is primarily suppressed. This form of response probably contributed to the allocation of a greater fraction of the dry matter to cassava tubers at low temperature, and to oil palm fruits in nutrient poor soils.

Concluding remarks

As for most other processes examined in earlier chapters, the control of period for which structures fill is better understood than the control of their rate of filling, as determined by the partition factor, p. It is known that the value of this factor in many species is largely independent of plant mass and dry matter production, is inherently larger for some structures than others and is restricted for some structures in extreme environments. The physiological processes controlling it over much of the range of conditions experienced by a species are still unclear.

Of the main environmental factors, the effects of temperature are again the best understood. Those of photoperiod are reasonably well understood and can be very large. Those of nutrients are not well understood, though seem to be small. The effects of drought are the largest, and least understood. It is usually unclear, for example, why genotypes differ so much in the rate at which a sink is produced during drought, and in the amount of dry matter that can be retranslocated from store.

References

1   Boerboom, 1978, gives this analysis, using cassava as an example. See Veltkamp, 1985, for comparison of cassava clones and effect of photoperiod.
2   Equations 5.1 to 5.3 can be used with most large structures. The growth of small units of dry matter, such as grains, is usually examined at a more detailed level. Goldworthy's paper, given in General reading, this chapter, includes an account of the physiology of grain growth. See also Reference 23.
3   Vegetative priority; Cock, Franklin, Sandoval and Juri, 1979 (cassava), Corley, 1973 (oil palm).
4   Ong, 1983b (pearl millet).
5   Satake, 1976 (rice), Yoshida, 1977 (rice).
6   Alvim, 1977 (cocoa), Squire, 1989b (pearl millet).
7   Altitude, temperature and p, Cooper and Law, 1978 (maize), Cooper, 1979 (maize), Chamberlin and Songchao Insomphun, 1982 (rice), Irikura, Cock and Kawano, 1979 (cassava), Williams, Wilson and Bate, 1975 (groundnut).
8   Photoperiod effects; Alagarswamy and Bidinger, 1985, Carberry and Campbell, 1985 (effect on optimum population), Craufurd and Bidinger, 1988.
9   Many examples of genotypic differences in the sensitivity of the reproductive sink to the timing of drought, for pearl millet (in relation to earliness, synchrony, etc.), are given by Mahalakshmi and Bidinger, 1985a, b, 1986, and Mahalakshmi, Bidinger and Raju, 1987.
10 Root/shoot weight ratios. Barley; Brown *et al.*, 1987 (change with time,

dryness, fertilizer), Gregory, Shepherd and Cooper, 1984 (dryness, fertilizer). Pearl millet; Azam-Ali, Gregory and Monteith, 1984a (population), Gregory and Squire, 1979 (dryness). Groundnut; Rao *et al.*, 1989 (indeterminate, population). Cassava; Connor, Cock and Parra, 1981 (fine roots and tubers, dryness).

11 For much retranslocation; Garrity, Sullivan and Watts, 1983 (sorghum, Fig. 5.7), Goldsworthy, 1970 (sorghum), Gregory and Squire, 1979 (pearl millet), Goldsworthy and Colgrove, 1974 (maize). For moderate or little retranslocation; Chamberlin and Songchao Insomphun, 1982 (rice; see also Reference 13), Sheldrake and Narayanan, 1979 (pigeon pea).

12 Reddy, Ghosh and Panda, 1986.

13 Coaldrake and Pearson, 1985b – a strong response to nitrogen in a pot experiment outdoors; but no response in a field experiment.

14 Murata, 1969, summarizes Japanese (and other) work on rice from the 1950s and 1960s. R. F. Chandler's article 'Plant morphology and stand geometry in relation to nitrogen', pp. 265–89 in *Physiological Aspects of Crop Yield* (see General reading) gives examples of effects of N on grain/straw ratios.

15 For example Lemcoff and Loomis, 1986.

16 Nambiar, Rego and Srinivasa Rao, 1986 (sorghum, and nodulating and non-nodulating groundnut).

17 Muchow, 1988b.

18 Pate and Minchin, 1980, sources of nutrients for grain in legumes.

19 Results of DeLoughery and Crookston, 1979, for maize at Minnesota, USA, are useful for exploring the effects of drought on W_v and p'. Data include cultivars of different duration grown at three sites differing in rainfall. In moist conditions, W_v is not systematically affected by time-to-flowering, but in dry, W_v is increased most in late-maturing types. This effect of rainfall is apparent for sorghum in Botswana, Rees, 1986a.

20 Ofori and Stern, 1987, sorghum/cowpea in Australia, see also Natarajan and Willey, 1980a, for population effects in sorghum/pigeon pea.

21 Natarajan and Willey, 1986 (millet or sorghum with groundnut).

22 Grain number and mass. Maize; Egharevba, 1977. Pigeon pea; Tayo, 1982, Venkataratnam and Green, 1979.

23 Effects of temperature, and other factors, on the rate and duration of grain filling are examined in a series of papers from Australia, for mainly temperate cereals; see Wardlaw, Sofield and Cartwright, 1980, for reference to earlier work. Also Ong, 1983b (pearl millet), Cooper, 1979 (maize, altitude).

24 Cowpea; Turk, Hall and Asbell, 1980a, c, Grantz and Hall, 1982. Pigeon pea; Sheldrake and Narayanan, 1979.

25 For example Yoshida and Parao, 1976 (for rice). For thermal growth rate, see Hawkins and Cooper, 1981 (maize in Kenya), Ong and Squire, 1984 (pearl millet, glasshouses, UK, and field at Hyderabad, India).

26 Oil palm; Broekmans, 1957, Corley, 1977 (review, and references to earlier work mainly in West Africa, by A. Beirnaert, L. D. Sparnaaij and others).

Further reading

On general matters of partition in relation to other processes in annuals, Evans and Wardlaw's review has probably not been surpassed. It also puts cellular processes (e.g. hormones, phloem transport) in context, but

Gifford and Evans give more emphasis to these aspects. Fischer, and Snyder and Carlson explore the influence of selection and breeding on partition. Cannell's review and analysis is of much wider significance than the 'tree' in the title suggests; many of the processes he describes are on a different scale of plant mass to those in the other reviews cited, and complement them. All the above include tropical species, but concentrate on temperate, and all attempt to bring out the principles of the subject, so the species and locations do not detract from their relevance. The essay by Goldsworthy is mainly on tropical species. This gives a comprehensive treatment of partition and other processes, and is a good general summary of the state of crop physiology in the early 1980s.

Cannell, M. G. R. (1984) Dry matter partitioning in tree crops. In: Cannell, M. G. R. and Jackson, J. E (eds) *Trees as Crop Plants*. Institute of Terrestrial Ecology, Monks Wood Experiment Station, Huntingdon PE17 2LS, UK, pp. 160–93.

Evans, L. T. and Wardlaw, I. F. (1976) Comparative physiology of grain yield in cereals. *Advances in Agronomy*, **28**, 302–59.

Fischer, R. A. (1981) Optimising the use of water and nitrogen through breeding of crops. *Plant and Soil*, **58**, 249–78.

Gifford, R. M. and Evans, L. T. (1981) Photosynthesis, carbon partitioning and yield. *Annual Review of Plant Physiology*, **32**, 485–509.

Goldsworthy, P. R. (1984) Crop growth and development: the reproductive phase. In: Goldsworthy, P. R. and Fisher, N. M (eds) *The Physiology of Tropical Field Crops*. Wiley Interscience, Chichester, pp. 163–212.

Snyder, F. W. and Carlson, G. E. (1984) Selecting for partitioning of photosynthetic products in crops. *Advances in Agronomy*, **37**, 47–72.

Chapter Six

Environmental and Physiological Control of Yield

This final chapter is an attempt to draw together the responses of crops to environment and cultivation. The methods of analysing yield and the limitations imposed on yield by the main climatic factors are first summarized. The central part of this chapter examines the effects of some of the main cultural factors – nutrients, plant population density and mixed cropping. Lastly, species are compared in terms of their main physiological attributes.

Analysis by principal limiting factors

Previous chapters examined the physiology of yield in terms of four types of processes – development, expansion, and production and partition of dry matter. All these factors determine the economic yield by influencing the fraction of the environmental resources that stands use during growth and the efficiency with which they use them. Although carbon dioxide from the atmosphere and nutrients from the soil comprise the yield, these substances have not been treated as the main limiting resources. The concentration of carbon dioxide in the air, although increasing slowly, varies too little with time and between sites to determine the differences in bulk dry matter production between stands. These differences are caused mainly by both variation in the solar irradiance, which provides the energy to incorporate the carbon dioxide, and variation in the water and nutrients available to roots. In the analysis, water has been given prominence over nutrients. This is partly because less is known of the effects of nutrients at the level of the stand, and partly because water is more continuously involved in the process of dry matter production. When leaves photosynthesize, water has to be transpired, but more nutrients are not necessarily taken up.

This section summarizes the main physiological attributes governing yield when solar radiation and water are limiting, and also the main effects on these attributes of temperature, photoperiod and saturation deficit.

Solar radiation and water

Throughout most of the humid tropics, and during good wet seasons in many drier parts, dry matter production depends on the amount of solar radiation a canopy intercepts and the efficiency with which it converts this radiation to dry matter. Solar radiation mainly affects the rate of net CO_2–exchange. The expansion and duration of the foliage are controlled by other factors such as temperature and photoperiod. Dry matter production by a canopy of a given size varies in proportion to the amount of incoming energy at a rate governed by the solar radiation conversion ratio (ε_s). Generally in these conditions, the partition factor (p) for the economic yield is not much affected by environmental factors and yield is a conservative fraction of the total

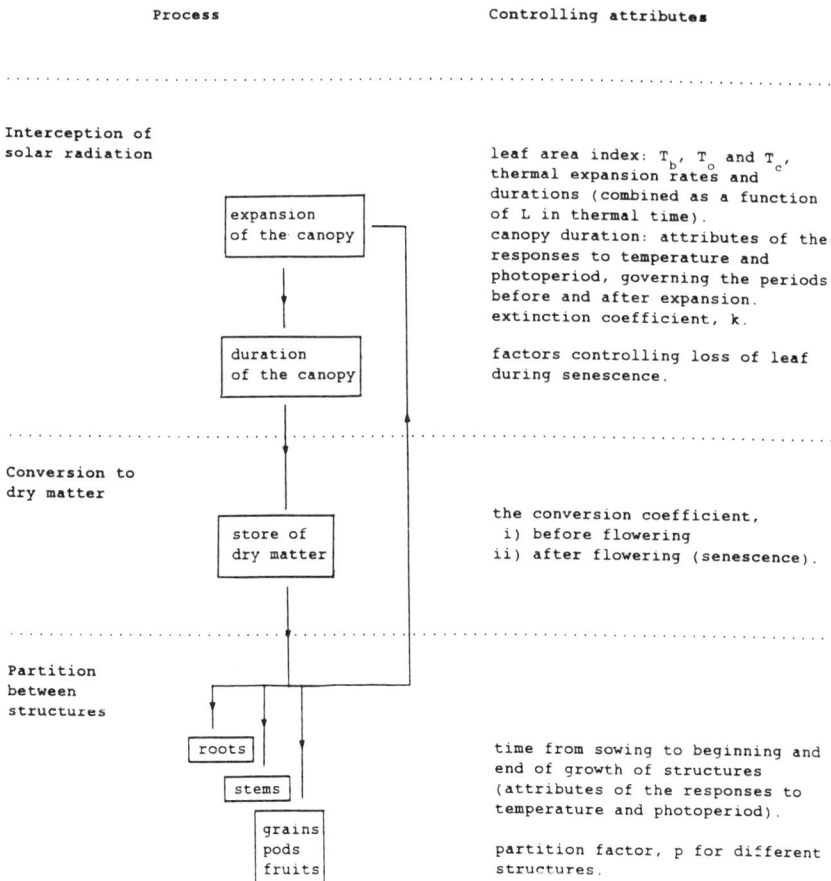

Process   Controlling attributes

Interception of
solar radiation

leaf area index: T_b, T_o and T_c,
thermal expansion rates and
durations (combined as a function
of L in thermal time).

expansion
of the canopy

canopy duration: attributes of the
responses to temperature and
photoperiod, governing the periods
before and after expansion.
extinction coefficient, k.

duration
of the canopy

factors controlling loss of leaf
during senescence.

Conversion to
dry matter

store of
dry matter

the conversion coefficient,
i) before flowering
ii) after flowering (senescence).

Partition
between
structures

roots

time from sowing to beginning and
end of growth of structures
(attributes of the responses to
temperature and photoperiod).

stems

grains
pods
fruits

partition factor, p for different
structures.

Fig. 6.1. Main attributes governing yield when solar radiation is limiting; i.e. the soil is moist and the saturation deficit < 1.5 kPa.

dry matter. Occasionally, the yield is sink-limited and much less dependent on the incoming or intercepted solar radiation. The main physiological processes and attributes governing yield are shown in Figure 6.1. Many attributes, such as the base temperature and the thermal durations, are conservative and are useful for modelling yield.

In most of the dry tropics where rainfall is sporadic and uncertain, and in irrigated agriculture, water is the limiting resource. Production depends on an amount of water extracted and on a dry matter/transpired water ratio (ε_w). The main attributes limiting extraction are those governing survival of individuals and expansion of the root system. The ratio, ε_w, is less affected than the conversion ratio for solar radiation by the state and amount of water or nutrients in the soil, but is strongly influenced by the saturation deficit. Partition factors are less stable in these conditions. Before any of the store of dry matter is allocated to yield, the stand must transpire for at least as long as it needs to set fruiting structures. Other important attributes affecting partition are the sensitivity of reproductive development to drought, and the amount of stored dry matter that can be moved between structures if dry matter production becomes very slow or ceases before development does. Figure 6.2 shows the main attributes. Several of them are far from conservative (e.g. root front velocity and effect on developmental durations), and governed by responses that are not understood.

Use of water: supply–limitation, demand–limitation

Either the radiation-limited or water-limited analysis can be used to examine the dry matter produced by a stand, though one is usually preferable. The product of the solar radiation, ΣS, the fractional interception, f, and the conversion ratio, ε_s, in equation 3.1, must be equal to the product of the transpired water, ΣE_t and the dry matter/water ratio, ε_w in equation 4.2, i.e.

$$\varepsilon_s f \Sigma S = \varepsilon_w \Sigma Et \qquad (6.1)$$

(This was the basis for estimating the demand for irrigation by crops in Chapter 4.)

When the soil is filled with water to near field capacity, the root system and the inflow (per unit root length) are large enough to supply water very rapidly to the foliage. Transpiration then depends on the incoming radiation and the size and photosynthetic activity of the canopy, and so is demand-limited. If a stand depletes the water in the upper layers of soil, transpiration depends on the rate at which the root system can extract water from deeper in the soil. Transpiration is then much less dependent on the energy available for evaporation, and is supply-limited. In this condition, the attributes ε_s and f both usually decrease to make the rate of loss from the canopy equal to the rate of supply[1].

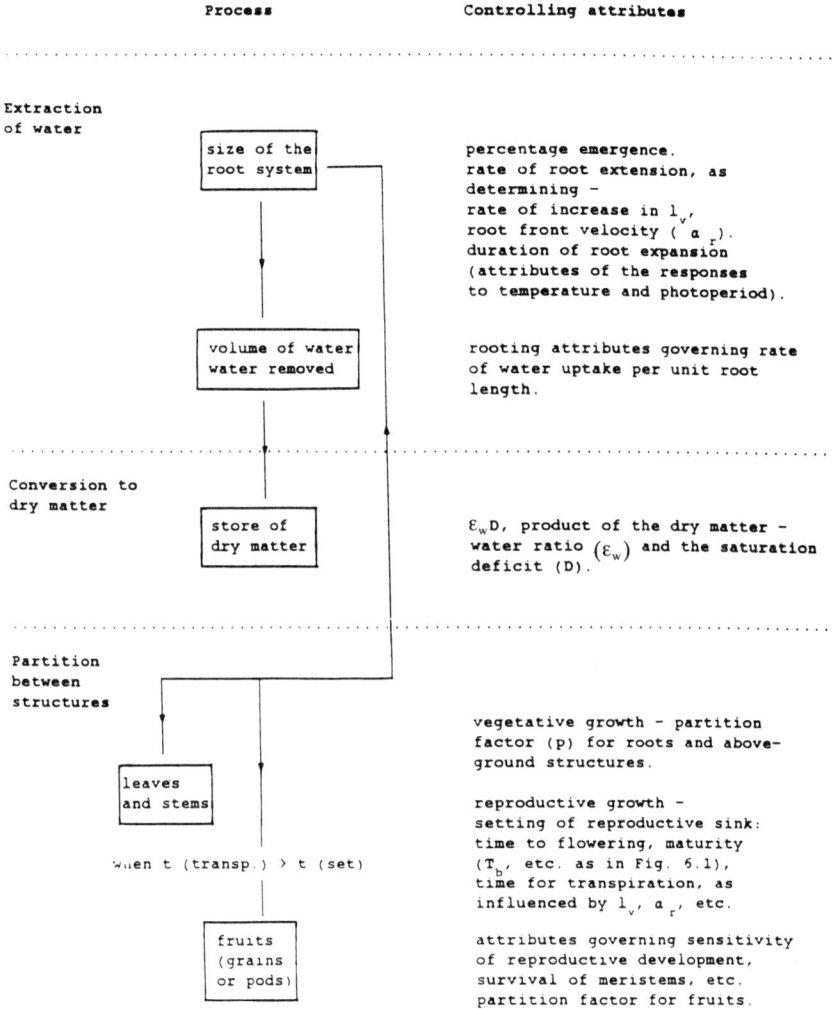

Process   **Controlling attributes**

Extraction of water

size of the root system — percentage emergence.
rate of root extension, as determining –
rate of increase in l_v,
root front velocity (a_r).
duration of root expansion
(attributes of the responses to temperature and photoperiod).

volume of water, water removed — rooting attributes governing rate of water uptake per unit root length.

Conversion to dry matter

store of dry matter — $\varepsilon_w D$, product of the dry matter – water ratio (ε_w) and the saturation deficit (D).

Partition between structures

vegetative growth – partition factor (p) for roots and above-ground structures.

leaves and stems

reproductive growth –
setting of reproductive sink:
time to flowering, maturity
(T_b, etc. as in Fig. 6.1),
when t (transp.) > t (set)   time for transpiration, as influenced by l_v, a_r, etc.

fruits (grains or pods) — attributes governing sensitivity of reproductive development, survival of meristems, etc. partition factor for fruits.

Fig. 6.2. Main attributes governing yield when water is limiting; i.e. when a stand is growing on a reserve of water and saturation deficit is typically greater than 2 kPa.

In the humid tropics, transpiration is demand-limited for much of the time. During hot, dry days, it might become temporarily supply-limited, if stomata close or leaves avoid the sun, even if the soil is moist. In the dry tropics, transpiration for stands growing mainly on a store of water is supply-limited for much of the time. It might become demand-limited during and immediately after rainy days. Between these extreme climates, transpiration probably changes many times during growth between supply-limited and demand-limited, depending on the frequency of rainfall and the size of

the canopy and root system. In these environments, and in irrigated agriculture, it is usually difficult to determine what primarily controls transpiration.

The influence of other climatic factors

The physiological attributes in both schemes are systematically affected by
other climatic factors; the main ones are temperature, photoperiod and dry
air.

Temperature

Temperature mainly governs the timing of processes, the rate of expansion
and survival of individuals or their reproductive structures; it has relatively
little effect on the conversion ratios. Temperature has most effect in the
analysis when solar radiation is limiting. In the analysis where soil water is
limiting, temperature affects yield mainly by determining the times to
flowering and later reproductive processes. The responses to temperature
are much affected by determinacy (Table 6.1).

Table 6.1. Some systematic effects of temperature in determinate and indeterminate
plants. Effects of increasing temperature in the range 20–30°C on some
morphological and physiological characters affecting yield. Time to harvest in the
indeterminate is assumed independent of temperature.

Character	Response	
	Determinate	Indeterminate
Duration: sowing to full canopy	decrease	decrease
Duration: full canopy to harvest	decrease	increase
Rate of expansion	increase	increase
Size of structures (including canopy)	little effect	increase
Extinction coefficient	little effect	little effect
Cumulative intercepted radiation	decrease	increase
Conversion coefficient	uncertain	uncertain
Dry matter production	decrease	increase
Potential number, reproductive units	little effect	increase
Actual number, reproductive units	decrease	variable*
Duration: growth of panicles/pods	decrease	increase
Mean weight of grain/pod	small decrease	uncertain
Partition factor, reproductive yield	little effect	variable*
Reproductive yield	decrease	variable*

*In types whose sink is not suppressed by temperature in this range, the actual sink,
  p and yield increase with rise in temperature; for those, e.g. groundnut, some
  cassava types, whose sink is suppressed, both p and yield decrease.

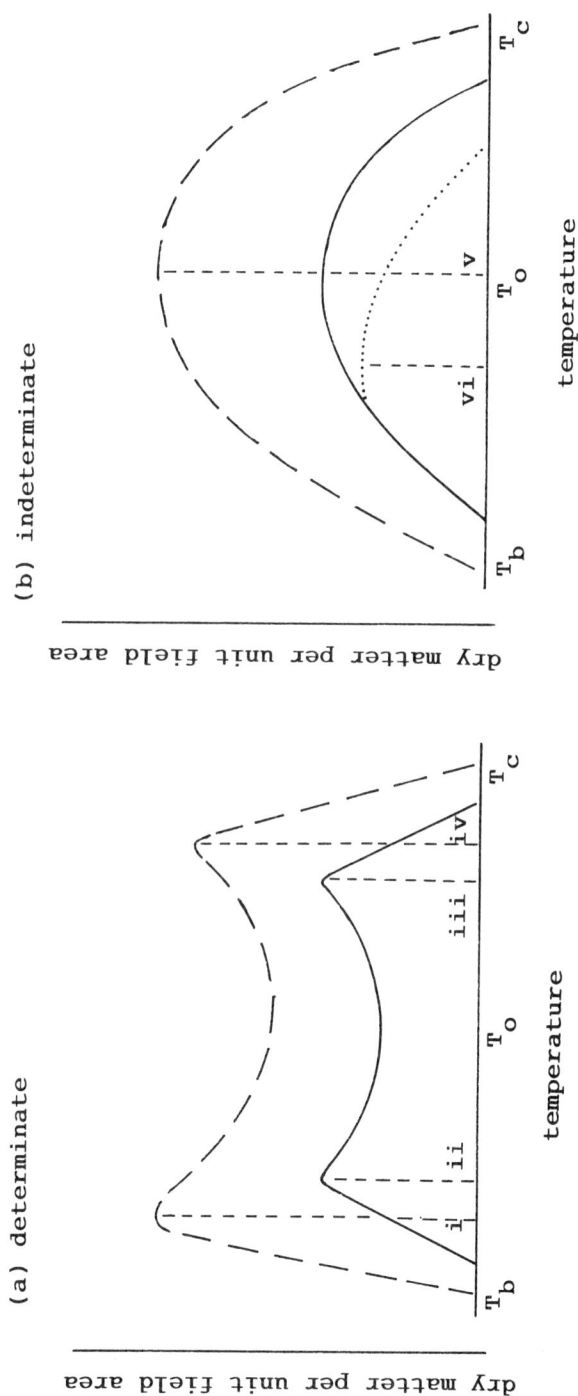

(a) determinate

(b) indeterminate

Fig. 6.3. Representation of responses to temperature in (a) determinate and (b) indeterminate genotypes. The dashed lines are total dry matter production. The continuous lines are yield for genotypes whose reproductive or storage sink is not sensitive to moderate temperature (sink-insensitive), i.e. most determinates and many indeterminates. The dotted line in (b) is yield for types whose reproductive or storage sink is suppressed at moderate temperature (sink- sensitive), e.g. groundnut, some cassavas. In both (a) and (b), T_b, T_o and T_c are the base, optimum and ceiling temperatures for developmental rate (equations 1.2 and 1.3). The other temperatures, indicated (i) to (vi), are different optima for dry matter. Determinate: the lower optimum for (i) total dry matter, and (ii) reproductive yield; the upper optima for (iii) total, and (iv) reproductive. Indeterminate: (v) the optimum for total dry matter in all types, and for reproductive or storage yield in sink-insensitive types; (vi) the optimum for reproductive or storage yield in sink-sensitive types.

In determinate cultivars (Fig. 6.3a), temperature governs dry matter much as it governs developmental duration. Over a wide intermediate range of temperature, the least dry matter is produced when the duration is shortest at T_o. As temperature decreases below T_o, production increases to a well defined lower optimum, which is usually several degrees above T_b. At temperatures below this lower optimum, the conversion ratio, ε_s, is reduced and few individuals survive. No dry matter is produced at or below T_b. The lower optimum for dry matter is typically 15–20°C. The response of reproductive yield is similar to that of dry matter, except the lower optimum temperature for yield might be higher than for total dry matter, since reproduction is usually restricted more than ε_s by low temperature. Comparable upper optima for total dry matter and yield are theoretically possible and indicated on Figure 6.3a, but might be irrelevant to the analysis when solar radiation is limiting, since dryness, rather than temperature, will usually determine rates of process in climates when temperature is mostly above T_o (i.e. typically > 35°C).

In indeterminate cultivars (Fig. 6.3b), the form of the response for total dry matter is similar to that for the rate of canopy expansion; the response therefore has a single optimum, similar to that for developmental and expansion rate (i.e. usually 30–35°C). Total dry matter decreases either side of this optimum as the duration of the effective canopy decreases, and might be very small near T_b and T_c, when temperature reduces survival and the conversion ratio. The form of the response of yield is commonly similar to that for dry matter. The optimum for yield however, might be much lower than for dry matter in those cultivars whose reproductive sink is very sensitive to moderately high temperature (e.g. groundnut), or whose canopy extension takes priority over economic yield (e.g. vegetatively vigorous cassava). In such genotypes, the optimum for yield can be 20–22°C, 8–10 degrees lower than for total dry matter.

The indeterminate plants have an optimum for dry matter close to the temperature of many tropical environments, whereas the determinates have an optimum that is much lower. Nevertheless, the advantage of C4 photosynthesis usually compensates, and among annuals, C4 determinates generally produce more dry matter than C3 indeterminates over most of the temperature range.

Photoperiod

Photoperiod affects mainly the duration of the vegetative phase of growth. As photoperiod rises from the critical to the ceiling (short-day plants), this phase lengthens and becomes an increasingly larger part of the crop's duration. Photoperiod has little effect on the conversion ratios, so (short-day) plants in long photoperiods produce more vegetative than reproductive dry matter. In environments where water is not limiting, photoperiod usually has little effect on the duration or rate of reproductive growth, and therefore

on reproductive yield, but reduces the fraction of the total dry matter allocated to yield (the harvest index). In environments where water is limiting
however, photoperiod can have a crucial effect determining whether developmental time matches the transpiration time. For example, a photoperiod that prolongs vegetative development might bring about a reduced
yield when there is a terminal drought, but an increased yield when there is
a mid-season drought.

Saturation deficit

The dryness of the air, represented by the saturation deficit (D), influences
the rate at which water is lost from plant tissue and moist surfaces. Saturation deficit mainly affects production of dry matter, though there are circumstances when D might affect partition of dry matter (through a tissue
water potential) independently of the state of water in the soil.

The effects of D on the relations between factors in equation 6.1 are very
complicated and not adequately understood[1]. In the radiation-limiting
analysis, D affects the rate of dry matter production by influencing both leaf
expansion and leaf conductance. The effects are inconsistent, but can be
large – sometimes reducing dry matter production by 20% per kPa increase
in D. In the water-limiting analysis, a large saturation deficit both reduces
the dry matter/transpired water ratio, ε_w (via $\varepsilon_w D$), and increases the rate of
evaporation from the soil surface; it might also affect the rate of transpira-
tion and thereby ΣE_t. The effect on ε_w is ubiquitous and very large: dry
matter production per unit water transpired is at least half in semi-arid environments (daily maximum D, 3–4 kPa) what it is in rainy seasons (1.5–2 kPa).
The saturation deficit is coupled with rainfall, but is little affected by
irrigation on a small scale, and should have an important effect restricting
the yield from irrigation.

The saturation deficit is important among the climatic factors for influencing yield mainly through the conversion ratios, particularly that for
transpired water. Temperature and photoperiod influence yield mainly
through developmental durations, the size of structures and partition
factors.

Physiological responses to cultural factors

Factors such as irrigation, fertilizer, population density and the composition
of the stand (e.g. intercropping) influence yield by affecting the rates and
durations of physiological processes. Of these factors, irrigation has been
the most widely and systematically studied, yet is practicable only in certain
localities. (Its physiological relations were summarized in Chapter 4.) Population density and fertilizer are important factors on every farm in the

tropics, but whereas population has been given adequate attention, at least in moist conditions, the physiological effects of fertilizer have rarely been studied and are insufficiently understood. Even less is known of how stand composition influences yield, despite its widespread importance especially in subsistence agriculture.

The account now concentrates on the physiological effects of fertilizer and population density. Responses described in preceding chapters are first summarized, then the analysis is taken further. The section ends with a summary of the physiological relations between the components of intercrops.

Nutrients as the limiting resource

The response of yield to nutrient dose is most commonly asymptotic. This response is the result of effects on several of the main processes leading to yield. Knowledge of these effects is patchy. Much of the information has been obtained in experiments on the main cereals.

Adding fertilizer to a nutrient-poor soil usually has little effect on the durations of developmental processes. It does however, increase the area of leaf and length of root. The additional leaf and root increase the amount of solar energy intercepted, or water transpired, during the growing season.

Adding fertilizer also increases the conversion ratio (ε_s) in some stands (probably through lowering c_i in equations 4.3 and 4.5); but it is uncertain if it does so in those that maintain a conservative nutrient content per unit leaf area. There are also indications, but little firm evidence, that individuals of stands at high density will sometimes produce so much leaf that the specific nutrient content will drop to values that limit ε_s. (The effects of nutrients on dry matter/transpired water ratio are also uncertain.)

Adding fertilizer reduces the fraction of assimilate allocated to the root systems, but generally has little effect on partition between vegetative and reproductive structures in the shoots. This might be because most plants can move nutrients between structures more easily than they can move stored carbohydrate.

The available evidence indicates that the major effects on yield of added nutrients are caused mainly by increases of leaf area and photosynthetic efficiency. Systematic and reproducible relations between nutrients and the physiological attributes in Figures 6.1 and 6.2 have been very slow to emerge. Therefore dry matter production is now examined with *nutrients* as the limiting resource[2]. In this analysis, which is comparable to those in Chapters 3 and 4 (equations 3.1 and 4.2), dry matter production is defined as the product of an amount of a specified nutrient taken up from the soil by the plant, and a dry matter/nutrient ratio (ε_n). The characteristics and control of uptake and ε_n are now summarized. Plants can be grouped in three categories: the C3 nitrogen-fixing legumes, the other C3 plants and the C4 cereals. Most of the information was obtained when nitrogen was the limiting resource.

Uptake of nutrients

The amount of nutrients accumulated in a plant during growth depends on the amount of nutrients available from different sources, and on the physiological processes that govern uptake and – in the case of nitrogen – fixation[3]. The uptake of nitrogen (N) by non-fixing annuals can be as little as 1 g m^{-2} from an infertile soil to as much as 30 g m^{-2} from a well-fertilized soil. Legumes can produce large amounts of nitrogen by fixation, and are much less dependent on soil N, though the amount of N in the shoots of legumes commonly increases when N fertilizer is added. Even without additional N, some legumes can accumulate as much as 20–30 g m^{-2} N (Fig. 6.4a).

The amount of nutrients taken from a fertilized soil is usually less than that given; and the fraction of the added nutrients taken up by the shoots

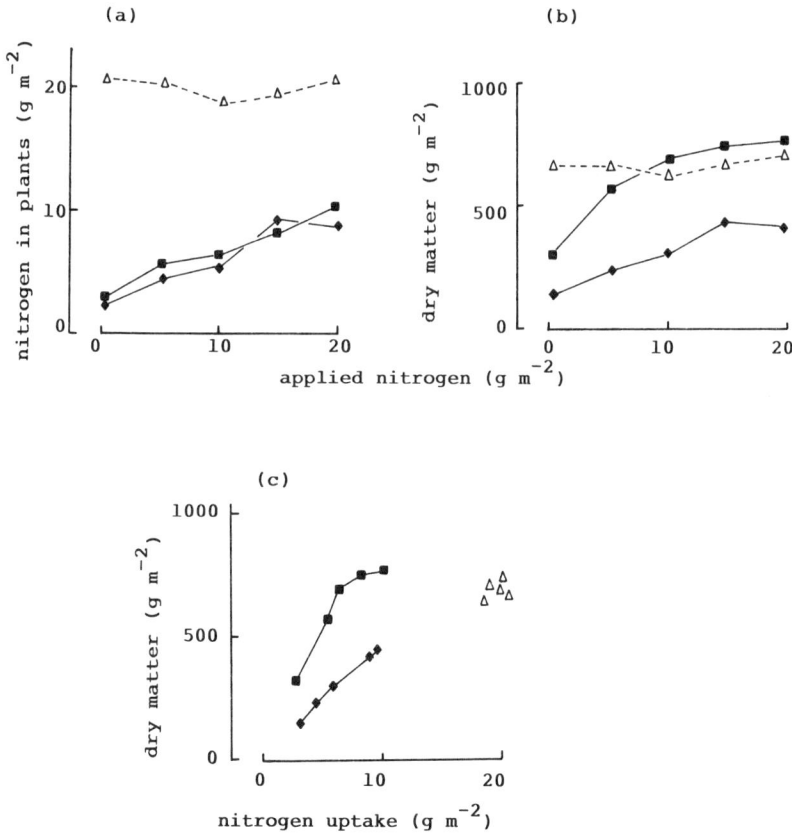

Fig. 6.4. **Relations between applied nitrogen, and the nitrogen and dry matter in plants at harvest.** For sorghum (■), a nodulating groundnut genotype (△), and a non-nodulating groundnut (♦), at Hyderabad, India. Data of Nambiar, Rego and Rao, 1986.

usually decreases the more nutrients that are added (Fig. 6.4a). At high doses (e.g. more than 20 g m^{-2} N), as little as 30% of that applied might be taken up by the shoots. (The additional amount in roots is seldom measured.) In moist conditions, nutrients lost in rainwater running off the soil surface account for some of the shortfall, but the losses must also depend much on the physiological (rooting) attributes of the genotype. Many other biological factors, such as the activity of mycorrhiza and other micro-organisms in the root zone, have been found to influence uptake. Their effects have rarely been compared with those of other factors in a thorough and complete physiological analysis.

Table 6.2 shows that the uptake of nutrients into the shoot system can differ a great deal between stands but does not vary systematically between the main groups. Nutrient uptake depends partly on the duration of the crop, as shown by the differences between the cereals in Table 6.2; but differences in the rate of uptake are sometimes as great as those of duration, as among the nodulating legumes. The rate of uptake is commonly faster in the C4 cereals than the C3 species, but does not seem to differ systematically between the non-fixing and fixing C3. Among the examples in Figure 6.5, the indeterminate C3 legumes and the root crop accumulated N throughout growth, whereas the rate of uptake decreased for maize after flowering, and ceased for pearl millet shortly after flowering. (The reason for this difference between the cereals is unknown: both had ample water.)

Many factors affect the rate of uptake. For example, it increases both with rise in population density and generally with increase in the solar radiation, but is sometimes smaller in dry than moist soil[5]. In many instances when nutrients are not in short supply, uptake seems determined less by the supply of nutrients in the soil than by the rate of growth of the stand. For example, the set of curves in Figure 6.5a is broadly similar to a corresponding set for fractional interception of solar radiation (not shown).

Uptake and rooting

Few studies in the field have linked the uptake of nutrients to rooting. Figure 6.6 summarizes the results of work with pearl millet in India, in which two treatments were compared, both of which were given the same fertilizer. Both treatments were established on irrigation; one was allowed to dry the soil without further irrigation and the other was irrigated from 30 days after sowing. In the dry treatment most root extension occurred before the soil had dried sufficiently to restrict extension; the root systems had a similar length and profile in the two treatments. The seasonal changes in the rates of uptake of N, P and K were broadly similar; only those for K are shown.

In the irrigated treatment, the rate of uptake increased during the first 30 d, though the inflow of nutrients (uptake per unit length of root per unit time) decreased sharply during this period. The maximum rate of uptake

Table 6.2. Examples of the amounts of nitrogen (N), phosphorus (P) and potassium (K) in well-fertilized stands at harvest. Per unit field area, in above-ground tissue except cassava for which the values include tubers.

Species	Duration (d)	Amount of nutrient (g m^{-2})			Source
		N	P	K	
Maize	154	22	–	–	Allison, 1984
Maize	125	27	–	–	Ofori and Stern, 1987
Sorghum	95	10	–	–	Nambiar *et al.*, 1986
Sorghum	77	7	1.9	14.0	Natarajan and Willey, 1980b
Pearl millet	82	9	1.0	10.0	Gregory, 1979
Cassava	365	32	3.7	24.0	Howeler and Cadavid, 1983
Groundnut:					
(non-nodulating)	138	9	–	–	Nambiar *et al.*, 1986
(nodulating)	138	25	–	–	Nambiar *et al.*, 1986
(nodulating)	118	21	–	–	Nambiar *et al.*, 1986
Cowpea	125	14	–	–	Ofori and Stern, 1987
Cowpea	63	16	–	–	Wien *et al.*, 1979
Pigeon pea	160	7	0.63	4.4	Natarajan and Willey, 1980b
Pigeon pea	165	9	0.58	–	Sheldrake and Narayanan, 1979
Pigeon pea	160	8	0.5	–	Sheldrake and Narayanan, 1979

(a)

(b)

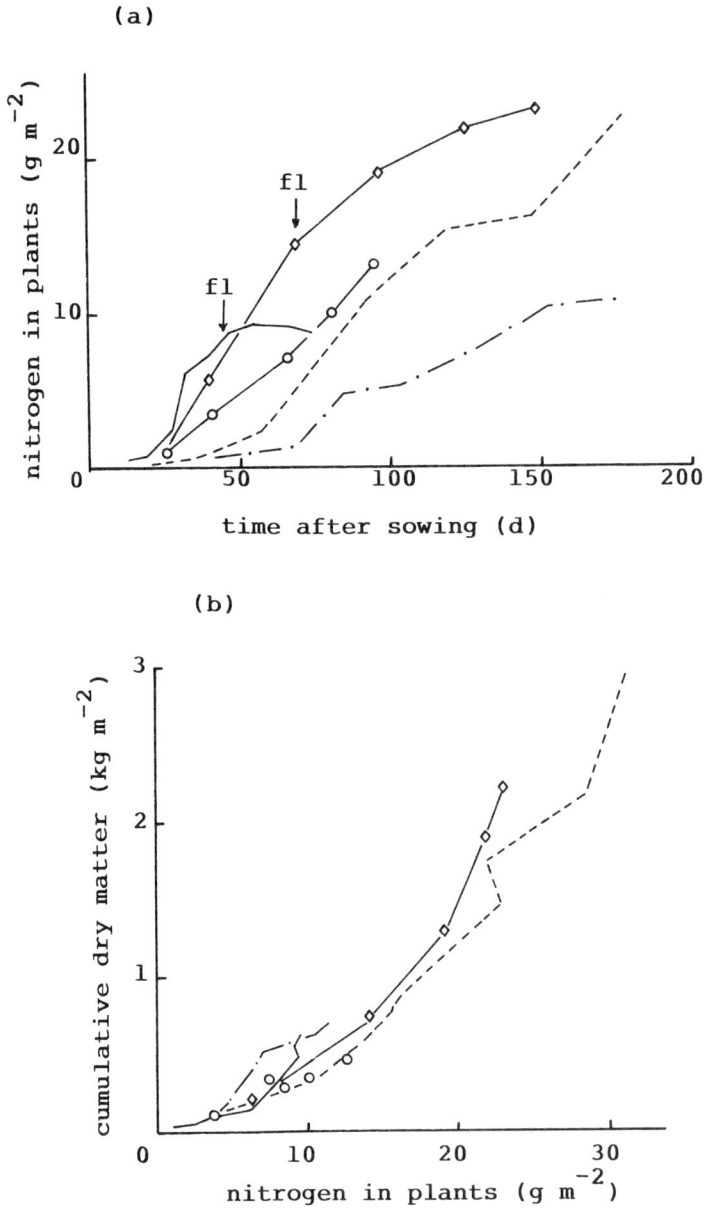

Fig. 6.5. Uptake of nitrogen and production of dry matter. Cumulative uptake during the growth of stands (a), and cumulative uptake in relation to cumulative dry matter production (b), for pearl millet (———), maize ($\Diamond$), groundnut ($\circ$), cassava (– – –) and pigeon pea (—·—). In (a), uptake by cassava continues, reaching 32 g m^{-2} after 12 months. For maize and millet, fl is flowering. Further information, Reference 8.

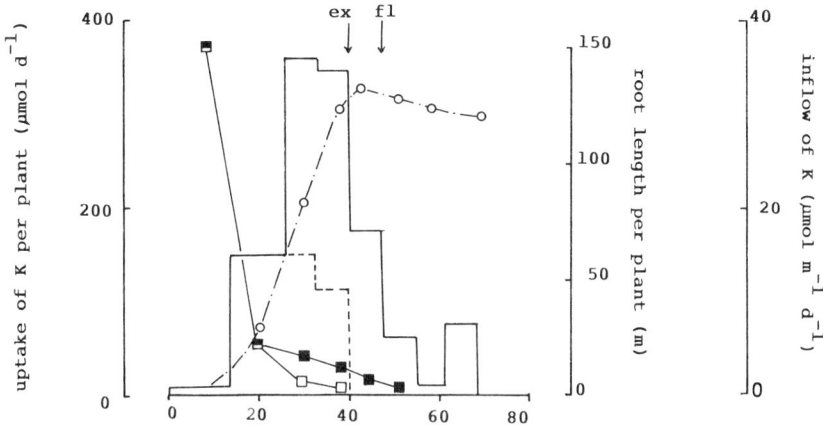

Fig. 6.6. Changes with time in the uptake of potassium and length of root. For pearl millet, either growing on a store of water (final irrigation on 13th day, 1 m deep alfisol), or irrigated periodically, at Hyderabad, India. Root length ($\circ$) was similar in both treatments. Rate of K uptake per plant (histograms): ———, irrigated; – – –, dry. Inflow of K per unit root length: ■, irrigated; □, dry. ex, end of canopy expansion; fl, flowering. From Gregory (1979) with additional information.

occurred around 35 d. After about 40 d from sowing, the rate of uptake decreased to very small values over a period of 20 d. The rates for all three nutrients decreased, but that of P decreased the most rapidly. In the dry treatment, uptake was reduced as early as 30 d, and was negligible after 40 d; again the uptake of P decreased earlier than that of N and K.

It is unclear whether the changes in nutrient uptake were governed primarily by the availability of nutrients in the soil, by the ability of roots to remove them from the soil or even by the requirement for nutrients by the shoots. The rate of uptake was largest well before the root length was maximum and declined when total root length was still roughly at its maximum. The changes in rate of uptake after about 20 d were therefore independent of total root length. Admittedly, the maximum possible inflow (per unit root length) might have changed with time (along with changes in the structure or physiology of the roots), but the root system after 40 d was still active in other respects. For example, it supplied water between 50 and 70 d at a rate equivalent to $0.5 \, E_o$ (2–3 mm d^{-1}). Moreover, the canopy in moist soil produced dry matter most rapidly between the end of expansion and the start of flowering – the period when nutrient uptake was about half its previous maximum.

In the dry treatment, the fall in nutrient uptake occurred while the whole root system was still expanding and happened about one week before the fall in transpiration and dry matter production. Most nutrients were probably in the surface layers of soil and the drying of these layers probably first restricted uptake. The continued transpiration and growth was maintained on water from deeper layers. Growth and nutrient uptake however, probably cease in this stand because the root system stopped expanding (for reasons not entirely clear, Chapter 4). The main problem of analysing this study and other similar ones is identifying whether uptake is governed by soil conditions or by the 'demand' for nutrients by the shoots.

The dry matter/nutrient ratio ε_n

When the cumulative uptake of nutrients in time in Figure 6.5a, is re-plotted in relation to cumulative dry matter production, the different curves become much closer, and there is little difference in the relation for leguminous and other species (Fig. 6.5b). This implies that in those nutrient-rich conditions, differences in dry matter production caused the differences in nutrient uptake (including N fixation) and were not the result of it.

The shape of the relation between cumulative nutrient uptake and cumulative dry matter production is similar for all the stands shown in Figure 6.5, and for many others. The dry matter/nutrient ratio, ε_n, increased with the passage of time and as more nutrients and dry matter were accumulated. (Conversely the percentage concentration of nutrients in the plant tissue decreased.) This seems the general response for all nutrients.

In indeterminate species, such as the cassava and pigeon pea shown in Figure 6.5, the decrease of ε_n in time is caused mainly by an increase in the rate of dry matter production as the canopy expands and intercepts more of the solar radiation: the rate of uptake changes little after about 50 d from sowing. Consequently, the relation between accumulated dry matter and accumulated N is almost exponential during expansion and linear thereafter. In the cereals, the rate of dry matter production similarly increases during expansion, but the rate of uptake decreases after flowering (but not usually as much as for the millet in Figure 6.5). Even when dry matter production decreases after flowering, the uptake also decreases all the more rapidly. In the cereals, the relation between cumulative dry matter and nutrients is commonly more or less exponential throughout growth. Regardless of the way in which dry matter and nutrients are related, ε_n increases in time (percentage concentration decreases) for most nutrients and in most species.

The ratio, ε_n, at harvest is influenced by factors other than the duration of the crop. High solar radiation in moist climates increases ε_n – by stimulating photosynthesis more than nutrient uptake; and drought after expansion decreases ε_n – by restricting photosynthesis when uptake has already ceased. More important for analysing the effects of fertilizer, the dose itself affects ε_n. In other than nodulating legumes, ε_n is largest when no nutrients are added and decreases as the dose is increased. The sorghum in Figure 6.4c responded in this way, but much greater responses have been measured. For example, ε_n for maize in Australia decreased from 190 g g^{-1} (N) to 90 g g^{-1} (N) as the fertilizer dose was increased, and responses of similar magnitude for sorghum have been found[6].

Differences in ε_n between species

The effects on ε_n of environment, and of the duration of growth, usually obscure inherent differences in ε_n between the main groups of species growing in different environments, as in Figure 6.5b. Yet when compared in the same environment, ε_n is consistently larger for C4 than for C3 species (Fig. 6.4). Figure 6.7 shows a set of paired stands – each of the pair grown at the same site and time and with little or no fertilizer (so values for the cereals should be about maximum). Each C4 accumulated less nitrogen than the corresponding nodulating legume but produced more dry matter than the

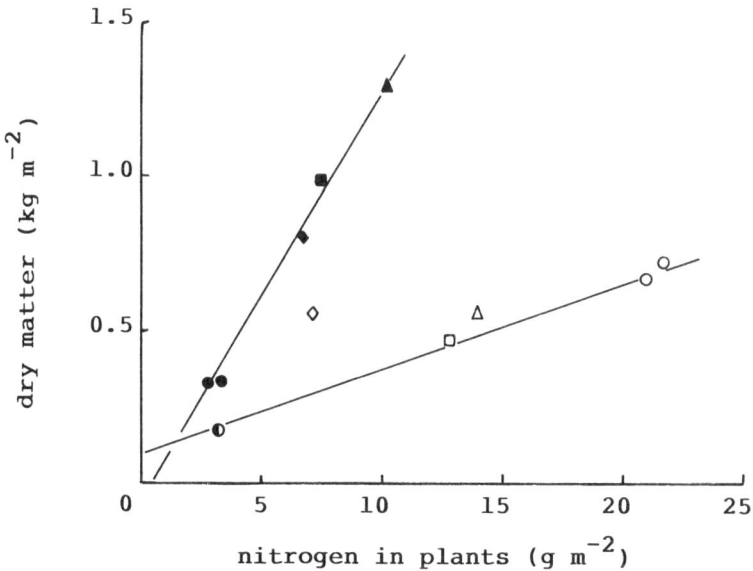

Fig. 6.7. Dry matter and nitrogen at harvest in C4 cereals and C3 legumes in nutrient-poor soils. Cereals (closed symbols) given no N except as indicated; legumes (open symbols) given no N. Natarajan and Willey, 1980a, b: ♦, sorghum (6 g m^{-2} N); ◊, pigeon pea. Reddy and Willey, 1981: ■, pearl millet (8 g m^{-2} N); □, groundnut. Nambiar *et al.* (1986); ●, sorghum (two genotypes), ○, groundnut (two genotypes), ◐, non-nodulating groundnut. Ofori and Stern (1987); ▲, sorghum, △, cowpea. At Hyderabad, India, except Ofori and Stern (Australia). Linear regressions (both axes g m^{-2}): cereals, y = $-50+133\pm9$ x (forced through origin, y = 127 ± 4 x); legumes, excluding pigeon pea, y = $107+28\pm3.1$x(through origin, y = 34 ± 1.8 x).

legume, except in two instances. The value of ε_n was larger for the C4 than the C3 of each pair. The mean value of ε_n was about 130 g g^{-1} (N) for the C4, and about 30 g g^{-1} (N) for the legumes (excluding pigeon pea, for which ε_n was unaccountably larger than for the other legumes).

The fourfold difference in ε_n between C4 and C3 species is much larger than the corresponding differences in the conversion ratios for solar radiation and transpired water. The greater energy content of legume pods than cereal grains accounts for some of this difference in ε_n[7]; but ε_n might have been small for the legumes simply because most of them grew for a longer period than the cereals, and accumulated large amounts of nitrogen (so compared with the C4 plants they were effectively growing in fertile soils, where ε_n for a species is small anyway). The non-nodulating groundnut in Figures 6.4 and 6.7 and legumes before podding possibly provide a better comparison with C4 species. The ε_n of the sorghum in Figure 6.7 was 1.8 times the value of the non-nodulating groundnut in the same experiment (averaged over all N treatments). Just before flowering, the corresponding factor was 1.7 for millet and groundnut and 2.3 for sorghum and pigeon pea. These factors are similar to those for the conversion ratios ε_s and ε_w.

Supply–limitation and demand–limitation of nutrient uptake

When nutrients are not limiting yield, as when large amounts of fertilizers are applied, the amount of nutrients taken up by the plants seems governed largely by the amount of dry matter they produce. This is also probably so for the nitrogen fixed by legumes when other nutrients are plentiful. In these circumstances, there is no fixed ratio of dry matter to nutrients; more nutrients will be taken up if more are available. These nutrients are not evenly distributed among the plant parts: they are more concentrated in the grains of cereals (e.g. 1.4–1.6% N by weight) and legumes (typically 3–5% N) than in vegetative structures. Among vegetative structures, the concentration is variable; in leaves, it can be as much as in some grains, but is usually small in stems (e.g. 0.4–0.8%). The fraction of the total nutrients taken up that is in the fruiting structures (the nutrient harvest index) is therefore greater than the harvest index for dry matter. For example, a harvest index of 0.4 would be about equivalent to a nitrogen harvest index of 0.6 in a cereal and 0.7 or more in a legume.

When nutrients are limiting yield, as among the cereals in Figure 6.7, the uptake seems to determine dry matter production. There is still not enough information to be certain that the value of the dry matter/nutrient ratio, common to the cereal stands in Figure 6.7, can be applied more widely. It is also uncertain how the ratio for nodulating legumes depends on other factors, including the scarcity of nutrients other than N, and how the N-fixation is (quantitatively) governed in such conditions. When nutrients are limiting, their concentration in plant parts is usually lower than when they are plentiful. The concentration is conserved in grain more than in other parts however. For example, the concentration of N in grain remained at 1.1% or above for the cereals examined in this chapter, and at 2.8% or above for the nodulating legumes. In contrast, the concentration in the rest of the plant sometimes decreased to almost half that in well-fertilized stands. This requirement for a high concentration of nutrients in grain must be an important factor limiting dry matter production in infertile soils[7].

Despite the consistent relations in Figure 6.7, the ratio ε_n is much less stable than ε_s and ε_w. Therefore, the analysis of dry matter production with nutrients as the limiting resource might prove to be less widely applicable than the analyses based on intercepted solar radiation and transpired water.

Plant population density

The principles governing the response of total dry matter production to plant population (N_p) are similar whether solar radiation or water is the main limiting factor. Population however, sometimes has different effects on the allocation of dry matter to yield in moist and dry conditions.

Effect on total dry matter production

At very low plant population densities (< 1 m^{-2} for most herbs) the roots and foliage of a plant rarely touch those of its neighbours. The resources used by the plant – the solar radiation intercepted or the water abstracted – are each a linear function of population. At a higher population, when some mingling occurs, more of the resources are used, but each plant 'competes' with its neighbours, and uses a smaller fraction of the resources. As N_p rises still further, there comes a point at which the stand leaves very little of the available resources. Over the range of N_p examined in most experiments, the relation between dry matter production and N_p is commonly an asymptotic curve, as in Figures 6.8a and 6.9a[11].

The steepness of the curve on which dry matter rises with N_p depends on the rate of leaf area expansion and on the extinction coefficient (or on the rate of root extension and other rooting attributes). In moist conditions, the maximum dry matter that the stand produces is determined by the incoming solar radiation, the duration of the stand, and the conversion ratio. In dry environments, the maximum generally depends on the amount of available water, and the conversion ratio. If the soil is deep and holds much water, the maximum dry matter increases as the rate and duration of root extension increase. If the soil is shallow (e.g. < 1 m), stands of most tropical species will extract a similar amount of water, irrespective of the rate and potential duration of extension.

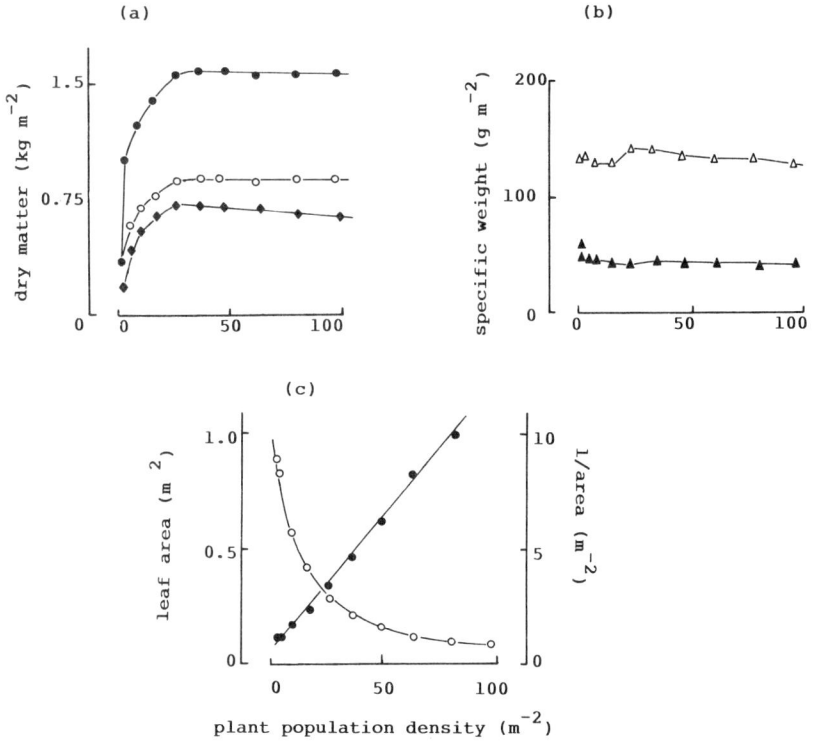

Fig. 6.8. **Some main attributes influenced by population density. I. Rice.** (a) Total above-ground dry matter ($\bullet$), vegetative dry matter ($\circ$), and grain dry matter ($\blacklozenge$), (b) vegetative dry matter per unit leaf area ($\triangle$) and leaf lamina dry matter per unit leaf area ($\blacktriangle$), (c) leaf area per plant ($\circ$) and the reciprocal of leaf area per plant ($\bullet$). Sampling times; (a) and (b) near flowering, (c) at maturity. Data of Akita, 1982a, b. In (c) the linear regression of l/area on plant population is y $= 0.82 + 0.11 \pm 0.021$ x.

In most experiments on population, in whatever conditions, the attributes controlling dry matter production (including the conversion ratios) are little affected by change of population. Even if the extinction coefficient decreases at high population (when leaves are held more upright), the resulting effect on interception of solar radiation is negligible, since most of the light is intercepted in any case. Dry matter production changes little with rise in N_p once most of the radiation has been intercepted. There are instances however, when the relation between dry matter and N_p shows an optimum above which dry matter production falls[12]. This seems to happen mainly in adverse conditions, and might in some cases, be spurious, as when more of the dry matter is allocated to roots in dry soils. It might occur if pests or disease spread more rapidly, and therefore reduce interception or conversion more in dense than sparse stands. It might also happen if the large

(a)

(b)

(c)

plant population density (m^{-2})

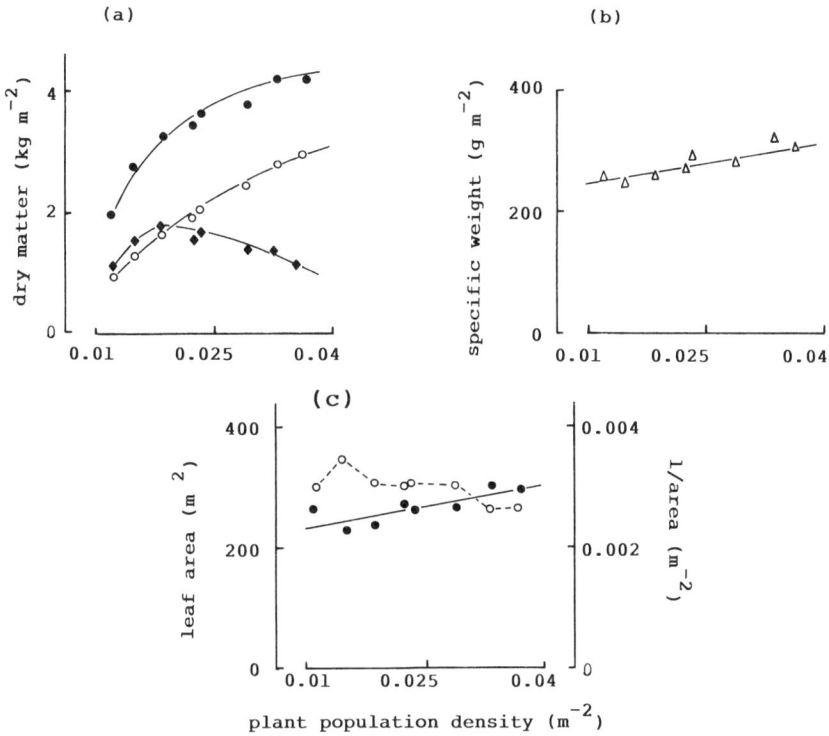

Fig. 6.9. Some main attributes influenced by population density. II. Oil palm. (a) Dry matter (carbohydrate equivalent) produced above-ground, in a year; whole plant (●), fronds and trunk (○) and fruit bunches (◆): (b) vegetative dry matter per unit leaf area, linear regression, y = 230+2390±550 x, P < 0.01: (c) leaf area per plant (○) and the reciprocal of leaf area per plant (●), regression of l/area on population, y = 0.0029+0.027±0.0089 x, P < 0.01. Data of Corley (1973) with dry matter adjusted to account for greater energy content of oil.

amount of leaf in dense stands 'dilutes' the nutrient content per unit area so much as to suppress the conversion coefficient. In many experiments, it is difficult to be certain whether any of these effects occur, simply because a loss of leaf before harvest (as commonly occurs in dense stands) would lead to an underestimate of production.

Partition and yield

In moist conditions, the general form of the response of yield to N_p is similar for most genotypes. The yield rises steeply with rise in population to an optimum, then decreases less steeply with further rise in N_p. The rates of rise and fall of yield, and the optimum population (at which yield is largest), are much influenced by genotypic traits. For example, the yield of the rice shown in Figure 6.8a increased very steeply per unit rise in N_p below the op-timum, but decreased only very slightly per unit rise above the optimum. In contrast, the yield of oil palm (Fig. 6.9a), increased much less steeply below the optimum and decreased more steeply above it. Moreover, the optimum population was much larger (by a factor of 10^3), and yield close to the maximum over a much wider range of N_p, in the rice than the oil palm.

Two main groups of attributes determine the characteristics of the response curve for yield. On the ascending part of the curve, the important attributes are those governing the amount of solar radiation each individual can intercept, such as the leaf area each can produce at low population, and the extinction coefficient. On the descending part, the attributes are the extent to which individuals can reduce their vegetative mass, to allow some of the dry matter they produce to be allocated to yield. In most circumstances, the critical attributes are those governing the effect of population on the area and mass of vegetative structures, e.g. parts (b) and (c) of Figures 6.8 and 6.9. It was shown in Chapter 2 that the leaf area of a plant is usually inversely proportional to population density, such that 1/area varies linearly with N_p, with a defined intercept and slope. The rice is a genotype with a 'plastic' vegetative mass: the leaf area decreases with increase of N_p, yet the vegetative dry matter per unit leaf area changes little. The oil palm is much more rigid in its response: leaf area hardly changes with N_p, and the dry matter per unit leaf area even increases slightly with rise of N_p.

For most other agricultural species, the response of 1/area to N_p lies between those of rice and oil palm. The vegetative mass/leaf area ratio is also conserved in many other species (but not always, though it is unclear what governs it). Absolute values of the mass/area ratio range between 200 and 300 g (vegetative dry matter) m^{-2} (leaf area) in maize, cassava and oil palm, to half this in slighter plants such as early pigeon pea. Despite this variation the relation between 1/area and N_p is the main discriminant of differences between genotypes in the response to population.

The importance of vegetative 'plasticity'

The effects of vegetative plasticity can be demonstrated by a simple model of yield in relation to population. Figure 6.10 shows the effects of changing the slope in the relation between 1/area and N_p. The intercept here is constant, so the curves in Figure 6.10a represent a series of plant forms which produce a similar leaf area when growing in isolation at low population, but which reduce leaf area (per plant) to different degrees as population rises. Other attributes are assumed to be unaffected by N_p: developmental timing, the conversion ratio and vegetative mass per unit leaf area. They are given arbitrary values, typical of a small indeterminate herb. When population has no effect on mean leaf area per plant, the response has a sharp optimum

(a)

(b)

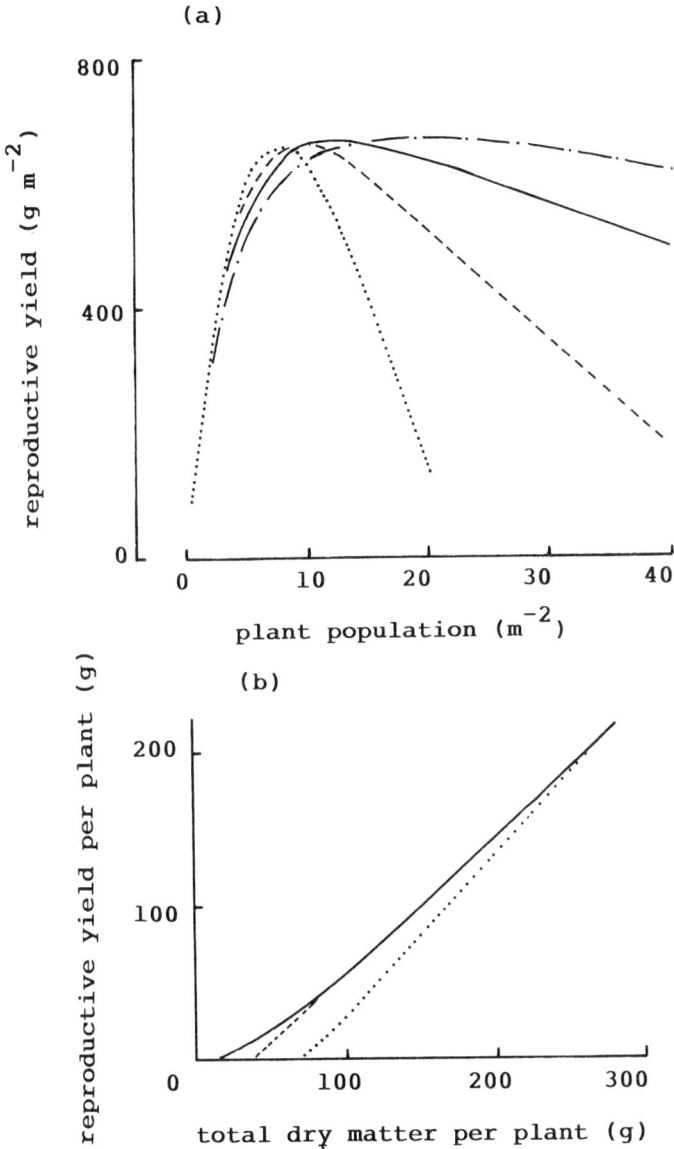

Fig. 6.10. Modelled effect of vegetative 'plasticity' on yield in relation to population density. In part (a), lines show reproductive yield when mean plant leaf area is constant at $0.47 \, m^2$ (. . . . . .), and when l/area (y) varies with population (x) as in $y = a+bx$, where b is 0.06 (– – – –), 0.11 (———) and 0.16 (— ·—), and a is constant at $2 \, m^{-2}$. Other constant attributes; solar radiation, $20 \, MJ \, m^{-2} \, d^{-1}$, extinction coefficient, 0.4, conversion ratio, $1.0 \, g \, MJ^{-1}$, crop duration, 80 d, vegetative dry mass per unit leaf area, $150 \, g \, m^{-2}$. In (b), relations are shown between reproductive and total dry mass, corresponding to two of the curves in (a). Text gives explanation.

(about 8 m^{-2}). Increasing the slope, which makes the leaf area per individual decrease more with rise of N_p, has two effects: the optimum N_p is larger (rising to 20 m^{-2}), since more plants are required to intercept most of the solar radiation but yield decreases much less steeply with rise of population above the optimum. The absolute value of yield at the optimum is independent of the slope but can be affected when other attributes are changed. For example, yield decreases as the vegetative mass per unit leaf area increases, and yield increases as the solar radiation or the conversion ratio increases.

Such simple models show that the optimum N_p of a genotype is determined by many environmental and physiological factors. Nevertheless, those attributes governing the vegetative characteristics should most strongly determine the response to N_p among the species and climates of the tropics. Characteristics such as a large leaf area per plant at low N_p, a weak response of leaf area to N_p and a large vegetative mass per unit leaf area, all decrease the optimum population[13]. These characters tend to cause an increase of W_v – the intercept in the relation between reproductive and total dry matter per plant (Chapter 5). This effect is shown in Figure 6.10b for two of the curves in Figure 6.10a. (At very high population and very small mass,

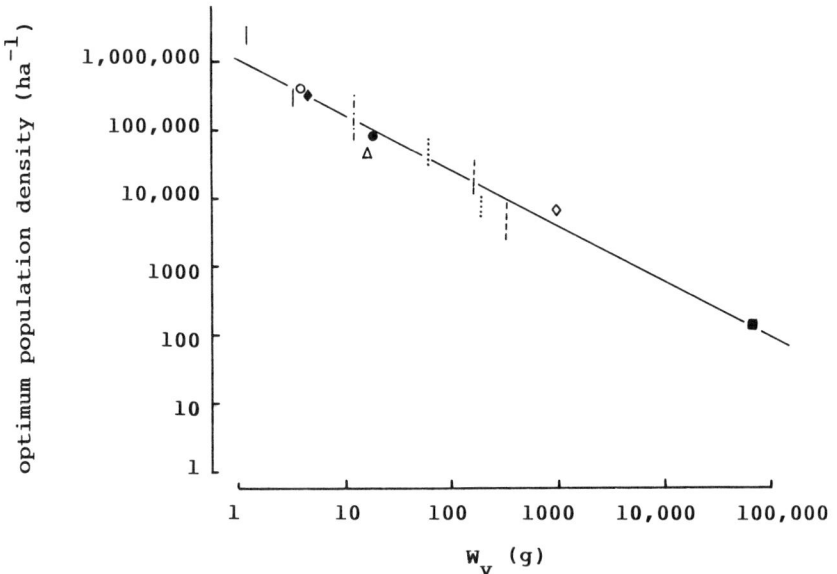

Fig. 6.11. **The relation between the optimum plant population density for yield (grain, fruits, tubers) and the apparent minimum vegetative dry mass, W_v.** Values of W_v are taken from Table 5.1. Vertical lines indicate a range of population over which yield was maximum, both axes on a logarithmic scale. Symbols: —, rice; –·–·, groundnut; ○, ..., pigeon pea; ◊, – – – –, cassava; ◆, pearl millet; ●, maize; △, sorghum; ■, oil palm. The line is the linear regression (using the mid-value where the optimum is a range) of log values: $y = 6 - 0.80 \pm 0.055\, x$.

the relation between reproductive and total mass is not linear, and W_v is determined by extrapolation from the linear part of the curve.)

In reality (as distinct from a model), the vegetative attributes are not constant in the face of change of N_p, but the extrapolated W_v is still an indicator of how much dry matter the plant allocates to vegetative as opposed to reproductive structures. For most tropical agricultural species, the optimum N_p decreases as the corresponding W_v increases (Figure 6.11, where the variables are expressed logarithmically for convenience). For several of the genotypes in Figure 6.11 sampling variation obscured the optimum population, and so a range of N_p is indicated over which yield was largest.

The effect of drought

In dry conditions, the effect of N_p on yield is influenced by additional factors. The main one is the length of time a stand can transpire in relation to the time-to-flowering and maturity. As N_p rises, there is less time for transpiration and more chance that reproduction will be impaired by low tissue turgor, or that there will be little current assimilate to fill the reproductive sink[14]. Accordingly, the optimum N_p is usually much lower in dry conditions than in moist, by as much as a factor of $10^{[15]}$. Certain attributes cause the yield to decrease much less steeply with increase in N_p above the optimum. Among these are root systems that permeate the available soil volume by means of fine roots, without producing many thick roots near the surface, and a capacity to retranslocate stored dry matter. Other advantageous attributes are early and synchronous reproduction for crops growing on a store of water, and late and desynchronized reproduction for those in a bimodal rainfall climate.

Even for genotypes adapted to dry conditions, the optimum density differs much between years depending on the amount and distribution of rainfall. For example, in Botswana[16], the optimum for sorghum over several years was 4 m^{-2}; though it was much lower in very dry years, when yield failed at high density, and higher in wet years when lower densities did not use all the available water. Although the manipulation of N_p offers a means of getting the best yield from a known supply of water, the choice of N_p is still very difficult when the amount of rain is uncertain.

Attributes affected by mixed cropping

One of the main conclusions of the physiological work on intercropping is how little the gross responses of a genotype are influenced by its being in a mixed stand. Stands of two species have attributes more or less in simple proportion to the relative populations of their components. Mixing seldom affects physiological attributes by as much as a factor of two, and commonly, effects are hardly significantly different.

Nevertheless, several responses have been found consistently in cereal-legume intercrops in moist conditions. Each plant of the taller species, usually a C4 cereal, intercepts more solar radiation when grown in an inter-crop, because it either produces more leaf, or holds leaves in a way that allows them to intercept more radiation per unit leaf area. Consequently, each plant produces more dry matter than one in a sole crop. If the cereal is a tillering type, more of the dry matter is allocated to tillers, but the partition between vegetative and reproductive structures is little affected. The indi-viduals of the shorter, usually C3, crop may intercept less radiation but use the solar radiation they intercept more efficiently – probably because the photosynthesis/light response is steeper at low intensities, under shade. Each individual produces about the same dry matter whether in sole or mixed crop, and partition of the dry matter is also little affected by the mix-ing. The mixed canopy as a whole uses solar energy much more efficiently than the legume, and sometimes slightly more than the cereal. The advant-age is such that 20–30% more land is usually needed to produce the same total yield if the components were grown as sole crops.

Generalizations are more difficult in dry conditions since there have been fewer experiments. Mixed cropping seems to have few systematic and reproducible effects on the main physiological attributes. An advantage in one species is commonly offset by a disadvantage in the other. Growing crops as mixed stands seems to get little more out of a poor environment than growing them separately. There might be other benefits such as con-venience at sowing, hindering the spread of pests and diseases, and reducing erosion, for example. Further work in nutrient-poor environments might reveal greater physiological advantages than have been shown so far.

Attributes for characterizing genotypes

In previous chapters, genotypes were compared in terms of the physiological attributes that govern their responses to environmental factors. Knowledge of these responses is uneven. Responses of development to temperature and photoperiod can be described in terms of a few fairly conservative attributes $(T_b, \hat{\Theta}_1, \text{etc.})$. Canopy extension can be similarly described, provided con-ditions are moist. But many responses can be described only partly quantit-atively. For example, extraction fronts of cereals can be said to descend at about 3 cm d^{-1} in loams, but it is not yet possible to present a function de-scribing the velocity of the front in terms of the mean pore size of the soil. The conversion ratio of a legume might, on a drying soil and in a saturation deficit of 3 kPa, be typically half the value in moist soil and a deficit of 1.5 kPa; but the ratio cannot be defined in terms of any index of the water–

environment, whether of the soil or atmosphere. Still other responses can be described only qualitatively. In the face of drought or nutrient stress, groundnut reduces leaf conductance and the conversion coefficient more than leaf area or light interception, but it can hardly be said by how much each is reduced, even less how the reductions are governed by the water or turgor potentials in any part of the leaf. Despite this unevenness, it is possible to move towards characterizing genotypes of the main tropical crops.

Some main attributes

A set of attributes that would characterize a genotype's broad responses to environment (at the level of the stand) is listed in Table 6.3. At present, no genotype can be described in terms of all of the attributes.

Several contrasting types are compared in Table 6.4 in terms of a smaller number of quantitative attributes. From Table 6.4, and from other analyses, it is possible to rank the main attributes approximately in order of their importance in determining differences in dry matter production and yield among the main tropical species.

1. *Survival* The ability of a plant to complete development at low or high temperatures, or during drought, is governed by many attributes, for which few quantitative relations have been found. There is much variation within each of the main groupings (e.g. C4 cereals, legumes, other C3). These attributes determine whether a genotype can be grown in adverse conditions, but have little influence on yield at moderate temperatures and when water is plentiful.

2. *Determinacy and perenniality* The degree to which most processes in a genotype are determinate or indeterminate influences both 'interception' of the limiting resource, and therefore the cumulative dry matter production (cf. a 75 day herb and a 30 year palm), and the partition factor for reproductive yield (the priority being given to vegetative structures in many indeterminates). Determinacy is special among all attributes in this list in influencing not only the size of a response to a variable, but also the direction of the response, particularly to temperature.

3. *Sink-sensitivity* The processes forming the reproductive sink are generally more sensitive to temperature and drought than are the processes governing dry matter production. Though genotypes differ very much in their sensitivity, the quantitative responses have been defined in only a few species, and little is known of the main physiological causes of sensitivity at the level of the stand.

Table 6.3. Some attributes for characterizing a genotype.

Development

T_b	base temperature.
T_o	optimum temperature.
T_c	ceiling temperature.
–	equation describing spread of times to germination or other events.
–	general description of degree of determinacy.
$\hat{\Theta}_1$	thermal duration (°Cd) below T_o for –
	germination,
	emergence (standard soil depth),
	leaf initiation,
	leaf expansion,
	tillering, branching,
	flowering processes not sensitive to photoperiod,
	filling of individual grain, pod, fruit,
	whole reproductive growth.
$\hat{\Theta}_2$	thermal duration (°Cd) above T_o for –
	germination,
	emergence,
	(very difficult for later processes).
–	general description of day-length sensitivity.
P_c	critical photoperiod.
P_{ce}	ceiling photoperiod.
P_b	base photoperiod.
φ	photoperiodic time.
–	where relevant, equations describing relation between base temperature and photoperiod.
–	in standard temperature and photoperiod (e.g. 28°C and at P_c) times to –
	emergence,
	flower initiation,
	flowering/grain, pod or fruit set,
	maturity (if determinate).

Expansion

T_b, etc.	usually as for development.
–	duration of expansion and of whole stand (usually controlled by development).
ρ_l	thermal expansion rate of leaf at representative ontogenetic position.
ρ_c	thermal expansion rate for whole canopy at optimum population density.
–	information on leaf senescence (qualitative).
	in soil of given physical properties:
α_r	root front velocity at optimum population for yield.
l_v	final rooting density at representative depths (e.g. 0.1–0.3 m, 1.0–1.2 m).

Interception of the resource

k	extinction coefficient.
–	propensity for leaf movement, rolling (qualitative).

Table 6.3. *contd.*

f	seasonal mean fractional interception of solar radiation, in moist conditions, standard temperature and photoperiod.
–	information on relations between rooting attributes and the uptake of soil water and nutrients (e.g. extractable water).

Dry matter ratios

ε_s	a. maximum conversion ratio for solar radiation (during vegetative growth, water and nutrients not limiting).
	b. maximum long-term value over vegetative and reproductive growth.
$\varepsilon_w D$	the product of dry matter/transpired water and the saturation deficit.
ε_n	the dry matter/nutrient ratio (specified nutrient, e.g. N) (a) when no added nutrients in a specified soil, and (b) nutrients not limiting.
–	description of ability to fix nitrogen.

Partition of dry matter (and nutrients)

–	intercept and slope for relation between l/(leaf area per plant) and population density.
–	vegetative dry matter per unit leaf area (and root length).
–	range of temperature over which reproduction is not impaired.
–	number of reproductive units per unit thermal growth rate.
–	maximum mean dry mass of reproductive units.
p	partition factor for main structures when water, nutrients not limiting.
W_v	intercept in relation between reproductive (or storage organ) and total dry matter per plant (from a population density experiment).
–	nutrient concentrations in main structures.

4. *Vegetative 'plasticity'* The extent to which individuals reduce their leaf area and vegetative dry mass strongly influences yield when total plant mass is reduced by adverse conditions or in dense stands. Generally, the reciprocal of leaf area or dry mass per plant increases linearly with population over a wide range (so the response can be described by an intercept and a slope).

5. *Nitrogen fixation* In nutrient-poor environments, the ability of nodulating legumes to fix atmospheric nitrogen gives them an advantage. They can produce twice the dry matter of C4 plants in nutrient-poor environments, and several times that of non-nodulating C3 plants.

6. *Photosynthetic efficiency* Genotypes with the C4 photosynthetic pathway consistently have conversion ratios about twice those with the C3 pathway. It is uncertain how much the conversion ratios vary among genotypes within each of these two groups. At least some genotypes of most of the species examined have long-term conversion ratios similar to the maximum for the respective group. This difference should decrease, and might eventually disappear, during the next century, if the concentration of carbon dioxide in the air continues to rise[9].

Table 6.4. Typical values of some characters governing production and partition for representative types.

Description	I annual determinate early C4	II annual determinate late C4	III annual determinate – C3 non-legume	IV annual ± indeterm. – C3 legume	V ± perennial indeterminate – C3 non-legume	VI perennial indeterminate – C3 non-legume
Dry matter production						
seasonal fractional interception	0.45	0.7	0.47–0.7	0.6	0.6	0.9
conversion ratio, maximum (g MJ^{-1})	2.5	2.5	1.3	1.3	1.0	1.0
$\varepsilon_w D$ (kPa g kg^{-1})	9.0		4.5		not known	
dry matter/N ratio, maximum (g g^{-1})	130.0	130.0	70.0	30.0	probably as for III	
Partition						
p for economic yield	0.9	0.9	0.9	0.6	0.7	0.7
W_v (g)*	2–4	2–4	1–4	3–15	100–1000	10^4
harvest index	0.45	0.3	0.45	0.4	0.6	0.6

*For definition of W_v, see Table 5.1.

7. *Attributes linking the extent of the 'intercepting' surface to vegetative mass*   Attributes such as the amount of dry matter that has to be expended per unit length of root or area of leaf, have an important role in determining partition of assimilate, and particularly in response to decreasing total plant mass (e.g. with rise in population). These mass/area (or mass/length) ratios are generally less variable than the mass, area or length themselves, but differ among the main species by up to a factor of about two.

8. *Developmental durations*   Tropical species differ little in the base and optimum temperature for development. Differences in developmental rate are related to differences in the thermal duration for processes, and in the attributes governing the response to photoperiod (the pre-inductive phase, the photoperiodic time, etc.). Such differences typically give rise to two- to threefold differences in dry matter production between genotypes of a species (water not limiting).

9. *Expansion rates*   The rate of expansion of individual leaves and plants can differ much between genotypes (cf. cereals and legumes) but generally the expansion rate of the whole canopy is governed more by the plant population. The rate of leaf expansion is not an important discriminant of production.

10. *Composition of the reproductive sink*   The proportion of protein and oil or fat, in the fruiting structures affects the efficiency of dry matter production, but is more important in determining the nutritional or commercial quality of the yield.

The relevance of some of these attributes to the two general characteristics of the growth and yield of a genotype is now examined: the fraction of the total dry matter that is harvested as yield and whether yield is mainly limited by the source of, or sink for, assimilate.

The harvest index

When the climate is moist and the temperature moderate, the harvest index depends more on attributes that govern the relative durations of processes, and the partition factors, than those that govern rates of production by the stand. For determinate genotypes, the duration of reproductive growth relative to total growth sets the upper limit for the harvest index. The ratio of the durations is about 0.5–0.6 for early cultivars of the cereals, and about 0.3 for later ones. Harvest indices are commonly less than these values because total production slows during the reproductive phase and dry matter is not always retranslocated to account for the shortfall.

The duration of reproductive growth relative to total growth is still important for indeterminate types. When the partition factor is stable, the harvest index increases with time after reproduction begins. The partition factor itself is determined by attributes such as vegetative plasticity and the leaf area to vegetative weight ratio, and also by factors governing the rate of dry matter production. In conditions of low production, the harvest index tends to be higher for types with a plastic vegetative (above ground) mass, such as indeterminate legumes, than for types with a more rigid vegetative mass, such as oil palm and cassava. This rigidity, however, can be an advantage in environments supporting high production. The highest harvest index of about 0.7 (based on energy) has been measured for stands of the rigid, indeterminate cassava and oil palm growing in very productive environments.

Source– or sink–limitation

As defined previously in this volume, the source is the supply of assimilate, both that produced when fruits and tubers are filling, and that stored in other parts of the plant. The sink is generally the number of sub-units within the yielding parts. Knowledge of whether the source or the sink limits the yield of a genotype can guide the strategy that should be adopted for improving the genotype by selection and breeding[17].

Source-limitation of reproductive yield is the most common condition among tropical species in a wide range of tropical environments. Many more fertilized inflorescences are available to form sinks than actually become sinks. Various regulatory mechanisms achieve a close match between sink and source. Cereals have the disadvantage of having to predict a match between the number of grains and the assimilate available during grain filling, but this number is usually an overestimate, so few grains fill to the maximum mass. Source-limitation is equally prevalent among determinate and indeterminate types. In some of the latter, a strong vegetative sink contributes to the condition.

Sink-limitation occurs most frequently in genotypes whose reproductive sink (or vegetative sink, if this is the yield) is sensitive to environment. Groundnut is commonly sink-limited for this reason, as are two vegetative crops – tea and rubber. The yield of most tea is habitually sink-limited because the young buds (which give rise to the yield) grow very slowly for most of their life and their rate of growth is sensitive to temperature and dry air[18]. Rubber is sink-limited because the lactifers delivering the latex become blocked in response to being severed.

It is not always possible to identify source- or sink-limitation. Experimentally, the limitation can sometimes be distinguished by manipulating the source or sink. For example, if the source is reduced (e.g. by removing leaves) and yield unaffected, there is sink-limitation; and if the number of reproductive sites is altered, but yield is unchanged, there is source-limitation[10]. Experiments of this type with oil palm and cassava have shown how the priority of the foliar sink over the reproductive or storage organ sink contributes much to source-limitation of yield. Similar studies with cocoa, pigeon pea and cowpea have shown that the reproductive sink can be diminished considerably without effect on yield.

Whether a genotype is source or sink-limited depends also on environment. Sink-limitation becomes more likely near the extremes of the climatic range of a genotype. This is because the reproductive processes are more sensitive to those extremes than vegetative ones, and ones determining rates of gas exchange. Sink limitation occurs, for example, when low or high temperature impairs pollination in cereals and when high temperature or soil capping reduces peg number in groundnut. Plants with protected inflorescences might be less affected by sink-limitation. Oil palm, for example, protects its developing inflorescences at an even temperature within the moist 'cabbage' of developing fronds at the top of the trunk. But even oil palm can be sink-limited when pollinating insects are absent.

The 'width' of the range of conditions in which the partition factor is reasonably stable differs much between, and even within species. Perhaps those cereals adapted to hot, dry conditions are stable over the widest ranges; while some of the legumes with sensitive and exposed reproductive structures are among the least stable. Taking the response to temperature, for example, the dryland cereals are stable over a range of at least 15°C, typically between mean temperatures of 20 and 35°C, whereas groundnut is stable probably only over 2–3 degrees between 20 and 25°C.

Main attributes: summary

The responses examined in this chapter show the importance of characters such as vegetative plasticity, sensitivity of the reproductive sink and retranslocation of stored assimilate. The main cause of differences in yield between many stands must lie in these attributes rather than in the ubiquitous differences attributable to photosynthetic efficiency and the ability to fix atmospheric nitrogen. Yet much less is known of the underlying control of, say, plasticity or sink-sensitivity, than of photosynthetic efficiency. Research should perhaps pay more attention to these intractable and unfashionable characters.

Concluding remarks

Physiological studies should benefit agriculture by improving the agronomist's ability to interpret and predict the effects of environment and cultivation on yield. The predictive value of physiology is particularly important for estimating the yield of a genotype in an environment where it has not actually been grown[18]. Predictions can be checked for existing combinations of weather and soil, albeit laboriously, by growing the genotype in a range of environments. Predictions, however, are the only way to assess how future climates might influence yields[9].

The physiological principles governing yield are most useful if they can be cast in a model of some form, whether as a series of simple relations scribbled on a piece of paper or a complex set of functions held in a computer. All models assume relations between environmental factors and physiological attributes and between different physiological attributes. The value and accuracy of the models – whatever their complexity – depends on how close the assumptions are to reality. The results of research show that the physiological principles governing many important processes are not well understood and, consequently, models are unable at present to predict with much certainty. The main virtue of crop modelling on any scale is to show which relations are least understood and which attributes might mainly govern a process[19].

Nevertheless, the physiological nature of a genotype can be reasonably well described in terms of a small number of attributes that govern development, expansion, production and partition. Most of these attributes can be measured in the field, and the most useful of them are conservative, in that they change little in response to environmental and cultural factors. Examples are the base temperature and thermal duration, the conversion ratio ε_s for solar radiation, the product of ε_w and D the saturation deficit and the partition factor.

Values can now be put on some of these attributes in a few specified environments for most of the main species, and in a wide range of environments for a few species (for example, Reference 20). This information is invaluable for comparing genotypes, environments and cultural systems and should be useful for assessing or estimating the effect on a regional scale of outstanding climatic events, or consistent changes in climate or soil conditions. For example, the implications from work in the laboratory that the rising CO_2 concentration in the atmosphere should increase the growth of C3 but not C4 species might be confirmed by future reports of conversion ratios of C3 plants higher than the current maxima. Those C3 species, such as cassava and oil palm, whose stands intercept much of the radiation received in a year and respond to many conditions by changing the conversion ratio, ε_s, might be the most sensitive indicators of change in CO_2 and other climatic factors, such as rainfall and saturation deficit.

The main deficiencies in knowledge, specific to the various processes, were briefly summarized at the end of each of Chapters 1 to 5. Work in these deficient areas is much more advanced in the laboratory than in the field. The range of responses of which genotypes are capable are far better understood than the responses they are most likely to take when growing in the field. Real progress will rely on coupling experiments in controlled conditions which fluctuate as in the natural environment and in which plants have unrestricted root systems, with experiments in the field, perhaps exploiting the effects of population density and altitude, among other agents, a little more.

There are, however, deficiencies of a more general nature. Few studies have examined the variation in the value of an attribute throughout a population (Chapter 1, Reference 1), or considered how this variation affects the yield of a stand in the field. The systematic effects of nutrients and the ability to fix nitrogen have yet to be examined in the same way as have those of, say, temperature or even saturation deficit. (This is not to discount the great volume of work on nutrients and nitrogen fixation, but to indicate there is still little known of the relations in the field between, for example, N-fixation and attributes such as the thermal duration for developmental events, the seasonal fractional interception, the conversion ratios, and partition factors.) The effects of pests and diseases are just beginning to be examined in terms of the physiological responses described here[21], but many other environmental constraints, including those of the soil, such as salinity and lack of oxygen, have yet to be investigated in detail.

The growth of physiological studies and analyses over the last 20 years has greatly depended on the development of environmental physics. This subject both stimulated a wave of better instrumentation[22], and taught that the most useful attributes of genotypes are those that define their physiology in terms of some element of the physical environment, rather than of themselves or their neighbours[23]. Physics and physiology have now become very much integrated at several research centres, and problems are tackled using both disciplines. The examples referred to in this volume show that progress can be made as rapidly in the small local research station as in the large international centres.

References and further reading

1   This analysis, and the role of saturation deficit, is taken further by Monteith, J. L. (1990) Steps in crop climatology. In: Unger, P. W., Jordan, W. R., Sneed, T. V. and Jensen, R. W. (eds) *Proceedings of the International Conference on Dryland Farming*, Texas A&M University Press (in press).
2   A useful introduction to the nitrogen balance is a chapter by Greenland, D. J. (1977) Contribution of microorganisms to the nitrogen status of tropical soils. In: Ayanabe, A. and Dart, P. J. (eds) *Biological Nitrogen Fixation in Farming Systems in the Tropics*, John Wiley and Sons, Chichester, pp. 13–25. Information on the role and concentrations of nutrients in plants and soil is summarized in Chapter 3, pp. 69–112, in *Russell's Soil Conditions and Plant Growth*, 11th edition (see General reading). Later chapters in the same volume deal in more detail with individual nutrients.
3   Various sections of *Russell* (11th edition) cover symbiotic nitrogen fixation: e.g. pp. 533–50, and pp. 643–51.
4   For nutrient uptake and rooting, pp. 144–63 in *Russell* (11th edition) is a useful summary.
5   Dryness; Gregory, 1979 (millet), Wien *et al.*, 1979 (cowpea).

6   Muchow and Davis, 1988 (maize), Ofori and Stern, 1987 (sorghum).

7   See papers by Penning de Vries (Chapter 3, Reference 10).

8   Stands in Figure 6.5. Maize; rainy season, fertilized, in Zimbabwe (Allison, 1969, 1984, at a population of 4.8 m^{-2}). Pearl millet; irrigated, fertilized, at Hyderabad, India (Gregory, 1979, Gregory and Squire, 1979). Cassava; fertilized, in Columbia (Howeler and Cadavid, 1983). Groundnut; rainy season, phosphate only, at Hyderabad (Reddy and Willey, 1981). Pigeon pea; rainy season, phosphate only, at Hyderabad (Sheldrake and Narayanan, 1979, cv. ICP-1 on vertisol, grown 1974).

9   Physiological modelling is still ill-equipped to predict the consequences of the rising CO_2 concentration and any associated change in climatic factors. Most laboratory studies indicate the growth of C4 species is unaffected by high CO_2, while the growth of C3 plants at tropical temperatures is increased by a factor of 1.3 on average. Surveys of responses to CO_2 are given by Kimball, B. A. (1983) Carbon dioxide and agricultural yield: an assemblage and analysis of 430 prior observations in *Agronomy Journal* **75,** 779–88; and Cure, J. D. and Acock, B. (1986) Crop responses to carbon dioxide in *Agricultural and Forest Meteorology* **38,** 127–45. For broad, though detailed, coverage (including the biochemical basis of the difference between C3 and C4 types), the book edited by E. R. Lemon, *CO₂ and Plants* (1983) American Association for the Advancement of Science, Washington, USA is recommended. Also, the section dealing with physiology in the review by Warrick, R. A., Gifford R. M. and Parry, M. L. CO_2, climatic change and agriculture. In: Bolin, B., Doos, B. R., Jager, J. and Warrick, R. A. (eds) *The Greenhouse Effect, Climate Change and Ecosystems,* SCOPE 29, John Wiley and Sons, Chichester, pp. 393–473. For a shorter summary of physiological effects, especially gas exchange, Morison, J. I. L. (in press). Plant growth in increased atmospheric CO_2, In: Ghazi, A. and Fantechi, R. (eds) *Current Issues in Climatic Research,* Reidel, Dordrecht.

10 Manipulating source and sink. Oil palm; summary in various chapters of *Oil Palm Research* (1976) Corley, R. H. V., Hardon, J. J. and Wood, B. J. (eds) Elsevier, Amsterdam, pp. 7–21, 77–86 and 291–8. Cocoa; Alvim, 1977. Cassava; Enyi, 1972a, b, Cock *et al.,* 1979, for summary, pp. 14–16 in Veltkamp, 1985. Pigeon pea; Sheldrake, Narayanan and Venkataratnam, 1979, for summary, see pp. 404–7, in chapter on Pigeon pea by A. R. Sheldrake (in Goldsworthy and Fisher, General reading). Cowpea; Ojehomon, 1970. Maize; Allison, 1984.

11 Some standard works on plant population. (a) Donald, C. M. (1963). Competition among crop and pasture plants. *Advances in Agronomy* **15,** 1–118. (b) Holliday, R. (1960). Plant population and crop yield. *Nature* **186,** 22–4. (c) Holliday, R. (1960) Plant population and crop yield. *Field Crop Abstracts* **13,** 159–67. (d) Willey, R. B. and Heath, S. B. (1969) The quantitative relationships between plant population and crop yield. *Advances in Agronomy* **21,** 281–321. For yield/density equations, see also; Counce, P. A. (1987) Asymptotic and parabolic yield and linear nutrient content responses to rice population density. *Agronomy Journal* **79,** 864–9.

12 Contrasting effects of population on dry matter production are shown by groundnut (Chapter 4, Reference 6) and pearl millet (Chapter 4, Reference 19). That of millet, showing a decline above an optimum, might have been spurious; only the tops were measured and more dry matter might have gone to roots at high N_p. The form of the asymptotic response of groundnut depended on the duration (100 d); given more time, the lower populations would probably have used as much water as the highest.

13 Cock *et al.* (1977) show the effects of different vegetative (shoot) attributes on optimum population and maximum yield for two cassava clones. Especially for vegetatively vigorous types, the population and canopy size giving optimum tuber yield are much smaller than those giving maximum dry matter production.
14 For example, in the experiment on groundnut in Reference 12, the highest population set least pods, since the stand used most of the water before reproduction began.
15 Among the variously droughted stands in Chapter 5, Reference 19, the optimum population was closely related to W_v.
16 Rees (1986a).
17 Physiology and plant breeding. (a) Blum, A. (1983) Genetic and physiological relationships in plant breeding for drought resistance. *Agricultural Water Management* **7**, 195–205. (b) Fischer, R. A. (1981) Optimising the use of water and nitrogen through breeding crops. *Plant and Soil* **58**, 249–78.
18 Hadfield (1968) inferred the yield of tea was source-limited, but see Tanton (1979) and Chapter 2, Reference 6 for more recent developments.
19 Breure's papers (main list) show the value of crop modelling at a reasonably practicable level. This approach can be studied in several of the publications from the Agricultural University Wageningen. e.g. Penning de Vries, F. W. T. and van Laar, H. H. (1982) *Simulation of plant growth and crop production. Simulation monograph.* Centre for Agricultural Publishing and Documentation, Wageningen, pp. 308. Also other books from this source.
20 Extensive surveys of f and ε_s; Squire, 1986 (oil palm, peninsular Malaysia), Hughes *et al.*, 1987 (chickpea, Syria).
21 A brief note in *Nature* **332**, p. 16 (1988) by G. Hughes, refers to several papers including P. E. Waggoner and R. D. Berger, Defoliation, disease and growth *Phytopathology* **77**, 393–8 (1987).
22 Books on instrumentation: *Instrumentation for Environmental Physiology* (1985) Marshall, B. and Woodward, F. I. (eds) Cambridge University Press, Cambridge. *Techniques in Bioproductivity and Photosynthesis* (2nd edition) (1985), Coombs, J., Hall, D. O., Long, S. P. and Scurlock, J. M. O. (eds) Pergamon Press, Oxford.
23 Witness the advantage of the conversion ratio, ε_s, over the misleading net assimilation rate!

References

Acevado, E., Fereres, E., Hsiao, T. C. and Henderson, D. W. (1979) Diurnal growth trends, water potential and osmotic adjustment of maize and sorghum leaves in the field. *Plant Physiology* **64,** 476–80.

Akinola, J. O. and Whiteman, P. C. (1974) Agronomic studies on pigeon pea (*Cajanus cajan* (L.) Millsp.) II. *Australian Journal of Agricultural Research* **26,** 57–66.

Akita, K. (1982a) Studies on competition and compensation of crop plants. IX. Effects of planting density on the characters of rice plant. *Scientific report of the Faculty of Agriculture, Kobe University, Japan* **15,** 5–10 (in Japanese).

Akita, K. (1982b) Studies on competition and compensation of crop plants. X. Effects of planting density on the growing organs of rice plant. *Scientific report of the Faculty of Agriculture, Kobe University, Japan* **15,** 11–16 (in Japanese).

Akita, K. (1982c) Studies on competition and compensation of crop plants. XI. Effects of planting density on the yield component in rice plant. *Scientific report of the Faculty of Agriculture, Kobe University, Japan* **15,** 17–21 (in Japanese).

Alagarswamy, G. and Bidinger, F. R. (1985) The influence of extended vegetative development and d_2 dwarfing gene in increasing grain number per panicle and grain yield in pearl millet. *Field Crops Research* **11,** 265–79.

Allison, J. C. S. (1969) Effect of plant population on the production and distribution of dry matter in maize. *Annals of Applied Biology* **63,** 135–44.

Allison, J. C. S. (1984) Aspects of nitrogen uptake and distribution in maize. *Annals of Applied Biology* **104,** 357–65.

Alvim, P. de T. (1977) Cacao. In: Alvim, P. de T. and Kozlowski, T. T. (eds) *Ecophysiology of Tropical Crops*. Academic Press, London, pp. 279–313.

Andrews, D. J. (1973) Effects of date of sowing on photosensitive Nigerian sorghums. *Experimental Agriculture* **9,** 337–46.

Angus, J. F., Cunningham, R. B., Moncur, M. W. and MacKenzie, D. H. (1981) Phasic development in field crops. I. Thermal response in the seedling phase. *Field Crops Research* **3,** 365–78.

Angus, J. F., Hasegawa, S., Hsiao, T. C., Liboon, S. P. and Zandstra, H. G. (1983) The water balance of post-monsoonal dryland crops. *Journal of Agricultural Science, Cambridge* **101,** 699–710.

Aresta, R. B. and Fukai, S. (1984) Effects of solar radiation on growth of cassava (*Manihot esculenta* Crantz). 2. Fibrous root length. *Field Crops Research* **9,** 361–71.

Azam-Ali, S. N. (1984) Environmental and physiological control of transpiration by groundnut crops. *Agricultural and Forest Meteorology* **33,** 129–40.

Azam-Ali, S. N., Gregory, P. J. and Monteith, J. L. (1984a) Effects of planting density on water use and productivity of pearl millet (*Pennisetum typhoides*) grown on stored water. 1. Growth of roots and shoots. *Experimental Agriculture* **20**, 203–14.

Azam-Ali, S. N., Gregory, P. J. and Monteith, J. L. (1984b) Effects of planting density on water use and productivity of pearl millet (*Pennisetum typhoides*) grown on stored water. 2. Water use, light interception and dry matter production. *Experimental Agriculture* **20**, 215–24.

Azam-Ali, S. N., Simmonds, L. P., Nageswara Rao, R. C. and Williams, J. H. (1989) Population, growth and water use of groundnut maintained on stored water. 3. Dry matter, water use and light interception. *Experimental Agriculture* **25**, 77–86.

Batchelor, C. H. (1984) The accuracy of evapotranspiration estimated with the FAO modified Penman equation. *Irrigation Science* **5**, 223–33.

Batchelor, C. H., Bell, J. P., Berthelot, B., Roberts, J. M., Robertson, C. A. and Soopramanien, G. C. (1985) First ratoon crop interim report. *MSIRI-IH Drip Irrigation Research Project*. Institute of Hydrology, Wallingford, Oxfordshire, UK.

Batchelor, C. H. and Roberts, J. M. (1983) Evaporation from the irrigation water, foliage and panicles of paddy rice in north-east Sri Lanka. *Agricultural Meteorology* **29**, 11–26.

Begg, J. E. and Burton, G. W. (1971) Comparative study of five genotypes of pearl millet under a range of photoperiods and temperatures. *Crop Science* **2**, 803–5.

Bell, M. J., Muchow, R. C. and Wilson, G. L. (1987) The effect of plant population on peanuts (*Arachis hypogaea*) in a monsoonal tropical environment. *Field Crops Research* **17**, 91–107.

Berry, J. and Bjorkman, O. (1980) Photosynthetic response and adaptation to temperature in higher plants. *Annual Review of Plant Physiology* **31**, 491–543.

Bierhuizen, J. F. and Slatyer, R. O. (1965) Effect of atmospheric concentration of water vapour and CO_2 determining transpiration–photosynthesis relationships. *Agricultural Meteorology* **2**, 259–70.

Black, T. A., Gardner, W. R. and Thurtell, G. W. (1969) The prediction of evaporation, drainage and soil water storage for a bare soil. *Soil Science Society of America Proceedings* **33**, 655–60.

Blacklow, W. M. (1972) Influence of temperature on germination and elongation of the radicle and shoot of corn (*Zea mays* L.). *Crop Science* **12**, 647–50.

Blum, A. and Ritchie, J. T. (1984) Effect of soil surface water content on sorghum root distribution in the soil. *Field Crops Research* **8**, 169–76.

Boerboom, W. J. (1978) A model of dry matter distribution in cassava (*Manihot esculenta* Crantz). *Netherlands Journal of Agricultural Science* **26**, 267–77.

Breure, C. J. (1988a) The effect of different planting densities on yield trends in oil palm. *Experimental Agriculture* **24**; 37–52.

Breure, C. J. (1988b) The effect of palm age and planting density on the partitioning of assimilate in oil palm (*Elaeis guineensis*). *Experimental Agriculture* **24**, 53–66.

Broekmans, A. F. M. (1957) Growth, flowering and yield of the oil palm in Nigeria. *Journal of the West African Institute for Oil Palm Research* **2**, 187–220.

Brown, S. C., Keatinge, J. D. H., Gregory, P. J. and Cooper, P. J. M. (1987) Effects of fertiliser, variety and location on barley production under rainfed conditions in Northern Syria. 1. Root and shoot growth. *Field Crops Research* **16**, 53–66.

Caddell, J. L. and Weibel, D. E. (1971) Effect of photoperiod and temperature on the development of sorghum. *Agronomy Journal* **63**, 799–803.

Callander, C. A. and Woodhead, T. (1981) Canopy conductance of estate tea in Kenya. *Agricultural Meteorology* **23**, 151–67.

Carberry, P. S. and Campbell, L. C. (1985) The growth and development of pearl millet as affected by photoperiod. *Field Crops Research* **11**, 207–17.

Carberry, P. S., Campbell, L. C. and Bidinger, F. R. (1985) The growth and development of pearl millet as affected by plant population. *Field Crops Research* **11**, 193–205.

Carr, M. K. V. (1985) Some effects of shelter on the yield and water-use of tea. In: Grace, J. (ed.) *Progress in Biometeorology (Volume 2)*. Swets and Zeitlinger B. V., Lisse, pp. 127–44.

Carr, M. K. V., Dale, M. O. and Stephens, W. (1987) Yield distribution in irrigated tea (*Camellia sinensis*) at two sites in Eastern Africa. *Experimental Agriculture* **23**, 75–85.

Chamberlin, R. J. and Songchao Insomphun (1982) Growth and development of upland rice at three altitudes in the highlands of northern Thailand. *Experimental Agriculture* **18**, 363–73.

Coaldrake, P. D. and Pearson, C. J. (1985a) Development and dry weight accumulation of pearl millet as affected by nitrogen supply. *Field Crops Research* **11**, 171–84.

Coaldrake, P. D. and Pearson, C. J. (1985b) Panicle differentiation and spikelet number related to size of panicle in *Pennisetum americanum*. *Journal of Experimental Botany* **36**, 833–40.

Coaldrake, P. D. and Pearson. C. J. (1986) Environmental influences on panicle differentiation and spikelet number of *Pennisetum americanum*. *Journal of Experimental Botany* **37**, 65–75.

Cock, J. H., Franklin, D., Sandoval, G. and Juri, P. (1979) The ideal cassava plant for maximum yield. *Crop Science* **19**, 271–9.

Cock, J. H., Wholey, D. and de la Casas, O. G. (1977) Effects of spacing on cassava (*Manihot esculenta*). *Experimental Agriculture* **13**, 289–99.

Coligado, M. C. and Brown, D. M. (1975) Response of corn (*Zea mays* L.) in the pre-tassel initiation period to temperature and photoperiod. *Agricultural Meteorology* **14**, 357–67.

Connor, D. J. and Cock, J. H. (1981) Response of cassava to water shortage 2. Canopy dynamics. *Field Crops Research* **4**, 285–96.

Connor, D. J., Cock, J. H. and Parra, G. E. (1981) Response of cassava to water shortage. 1. Growth and yield. *Field Crops Research* **4**, 181–200.

Connor, D. J. and Palta, J. (1981) Response of cassava to water shortage 3. Stomatal control of plant water status. *Field Crops Research* **4**, 297–311.

Constable, G. A. and Rawson, H. M. (1980) Carbon production and utilization in cotton: inferences from a carbon budget. *Australian Journal of Plant Physiology* **7**, 539–53.

Cooper, P. J. M. (1979) The association between altitude, environmental variables, maize growth and yields in Kenya. *Journal of Agricultural Science, Cambridge* **93**, 635–49.

Cooper, P. J. M., Gregory, P. J., Keatinge, J. D. H. and Brown, S. C. (1987) Effects of fertilizer, variety and location on barley production under rainfed conditions in Northern Syria. 2. Soil water dynamics and crop water use. *Field Crops Research* **16**, 67–84.

Cooper, P. J. M., Gregory, P. J., Tully, D. and Harris, H. C. (1987) Improving water use efficiency of annual crops in the rainfed farming systems of West Asia and North Africa. *Experimental Agriculture* **23**, 113–58.

Cooper, P. J. M. and Law, R. (1978) Environmental and physiological studies of maize. *Maize Agronomy Research Project, Final Report:* **Part 3, Vol. 1.** Kenya Ministry of Agriculture, U.K. Overseas Development Administration (Eland House, Stag Place, London).

Corley, R. H. V. (1973) Effects of plant density on growth and yield of oil palm. *Experimental Agriculture* **9**, 169–80.

Corley, R. H. V. (1977) Oil palm yield components and yield cycles. In: Earp, D. A. and Newall, E. (eds) *International Developments in Oil Palm*, Incorporated Society of Planters, Kuala Lumpar, Malaysia, pp. 116–29.

Corley, R. V. H. (1986) Yield potentials of plantation crops. In: *Potassium in the Agricultural Systems of the Humid Tropics*. Proceedings of the 19th Colloquium, International Potash Institute, Worblaufen-Bern, pp. 61–80 .

Corley, R. H. V., Gray, B. S. and Ng, S. K. (1971) Productivity of the oil palm (*Elaeis guineensis* Jacq.) in Malaysia. *Experimental Agriculture* **7**, 129–36.

Covell, S., Ellis, R. H., Roberts, E. H. and Summerfield, R. J. (1986) The influence of temperature on seed germination rate in grain legumes. 1. A comparison of chickpea, lentil, soyabean and cowpea at constant temperatures. *Journal of Experimental Botany* **37**, 705–15.

Craufurd, P. Q. and Bidinger, F. R. (1988) Effect of the duration of the vegetative phase on crop growth, development, and yield in two contrasting pearl millet hybrids. *Journal of Agricultural Science, Cambridge* **110**, 71–9.

Curtis, D. L. (1968) The relation between yield and date of heading of Nigerian sorghums. *Experimental Agriculture* **4**, 93–101.

DeLoughery, R. and Crookston, K. R. (1979) Harvest index of corn affected by population density, maturity rating and environment. *Agronomy Journal* **71**, 577–80.

Doorenbos, J. and Pruitt, W. O. (1975) (Revised 1977). Crop water requirements. *Irrigation and Drainage Paper* **No. 24**, FAO, Rome.

Dow El-Madina, I. M. and Hall, A. E. (1986) Flowering of contrasting cowpea (*Vigna unguiculata* (L.) Walp.) genotypes under different temperatures and photoperiods. *Field Crops Research* **14**, 87–104.

Eckardt, F. E. (1987) The controlled-environment plant-chamber technique for CO_2–exchange measurements in tree and forest research. In: Sethuraj, M. R. and Raghavendra, A. S. (eds) *Tree ·Crop Physiology*. Elsevier, Amsterdam, pp. 3–27.

Egharevba, P. N. (1977) The effect of planting configurations and population density on light interception, growth and production of maize in the lowland tropics. *Nigerian Journal of Science* **11**, 149–65.

Ellis, R. H., Covell, S., Roberts, E. H. and Summerfield, R. J. (1986) The influence of temperature on seed germination rate in grain legumes. II. Intraspecific variation in chickpea (*Cicer arietinum* L.). *Journal of Experimental Botany* **37**, 1503–15.

Ellis, R. H., Simon, G. and Covell, S. (1987) The influence of temperature on seed germination rate in grain legumes. III. A comparison of five faba bean genotypes at constant temperatures using a new screening method. *Journal of Experimental Botany* **38**, 1033–43.

El-Sharkawy, M. A. and Cock, J. H. (1984) Water use efficiency of cassava. 1. Effects of air humidity and water stress on stomatal conductance and gas exchange. *Crop Science* **24**, 497–501.

El-Sharkawy, M. A., Cock, J. H. and Held, A. (1984) Water use efficiency of cassava. 2. Differing sensitivity of stomata to air humidity in cassava and other warm-climate species. *Crop Science* **24**, 503–7.

Enyi, B. A. C. (1972a) Effects of spacing on growth, development and yield of single and multishoot plants of cassava (*Manihot esculenta* Crantz). 1. Root tuber yield and attributes. *East African Agricultural and Forestry Journal* **38**, 23–6.

Enyi, B. A. C. (1972b) Effects of spacing on growth, development and yield of single and multishoot plants of cassava (*Manihot esculenta* Crantz). 2. Physiological factors. *East African Agricultural and Forestry Journal* **38**, 27–34.

Enyi, B. A. C. (1973) Growth rates of three cassava varieties (*Manihot esculenta* Crantz) under varying population densities. *Journal of Agricultural Science, Cambridge* **81**, 15–28.

Evans, L. T., Visperas, R. M. and Vergara, B. S. (1984) Morphological and physiological changes among rice varieties used in the Philippines over the last seventy years. *Field Crops Research* **8**, 105–24.

Feddes, R. A. (1972) Effects of water and heat on seedling emergence. *Journal of Hydrology* **16**, 341–59.

Fischer, K. S. and Palmer, A. F. E. (1984) Tropical maize. In: Goldsworthy, P. R. and Fisher, N. M. (eds) *The Physiology of Tropical Field Crops*. John Wiley and Sons, Chichester, pp. 213–48.

Frere, M. and Popov, G. F. (1979) Agrometeorological crop monitoring and forecasting. *Plant Production and Protection Paper* **No. 17**, FAO, Rome.

Fukai, S., Alcoy, A. B., Llamelo, A. B. and Patterson, R. D. (1984) Effects of solar radiation on growth of cassava (*Manihot esculenta* Crantz) 1. Canopy development and dry matter growth. *Field Crops Research* **9**, 347–60.

Garcia-Huidobro, J., Monteith, J. L. and Squire, G. R. (1982a) Time, temperature and germination of pearl millet (*Pennisetum typhoides* S. & H.). 1. Constant temperature. *Journal of Experimental Botany* **33**, 288–96.

Garcia-Huidobro, J., Monteith, J. L. and Squire, G. R. (1982b) Time, temperature and germination of pearl millet (*Pennisetum typhoides* S. & H.). 2. Alternating temperature. *Journal of Experimental Botany* **33**, 297–302.

Garcia-Huidobro, J., Monteith, J. L. and Squire, G. R. (1985) Time, temperature and germination of pearl millet (*Pennisetum typhoides* S. & H.). 3. Inhibition of germination by short exposure to high temperature. *Journal of Experimental Botany* **36**, 338–43.

Garrity, D. P., Sullivan, C. Y. and Watts, D. G. (1983) Moisture deficits and grain sorghum performance: drought stress conditioning. *Agronomy Journal* **75**, 997–1004.

Goldsworthy, P. R. (1970) The sources of assimilate for grain development in tall and short sorghum. *Journal of Agricultural Science, Cambridge* **74**, 523–31.

Goldsworthy, P. R. and Colegrove, M. (1974) Growth and yield of highland maize in Mexico. *Journal of Agricultural Science, Cambridge* **83**, 213–21.

Grantz, D. A. and Hall, A. E. (1982) Earliness of an indeterminate crop *Vigna unguiculata* (L.) Walp., as affected by drought, temperature and plant density. *Australian Journal of Agricultural Research* **33**, 531–40.

Gregory, P. J. (1979) Uptake of N, P and K in irrigated and unirrigated pearl millet (*Pennisetum typhoides*). *Experimental Agriculture* **15**, 217–23.

Gregory, P. J. (1983) Response to temperature in a stand of pearl millet (*Pennisetum typhoides* S. & H.). 3. Root development. *Journal of Experimental Botany* **34**, 744–56.

Gregory, P. J. (1986) Response to temperature in a stand of pearl millet (*Pennisetum typhoides* S. & H.). 8. Root growth. *Journal of Experimental Botany* **37**, 379–88.

Gregory, P. J. (1987) Development and growth of root systems in plant communities. In: Gregory, P. J., Lake, J. V. and Rose, D. (eds) *Root Development and Function*. Cambridge University Press, pp. 147–66.

Gregory, P. J. and Reddy, M. S. (1982) Root growth in an intercrop of pearl millet/groundnut. *Field Crops Research* **5**, 241–52.

Gregory, P. J., Shepherd, K. D. and Cooper, P. J. (1984) Effects of fertiliser on root growth and water use of barley in Northern Syria. *Journal of Agricultural Science, Cambridge* **103**, 429–38.

Gregory, P. J. and Squire, G. R. (1979) Irrigation effects on roots and shoots of pearl millet (*Pennisetum typhoides* S. & H.). *Experimental Agriculture* **15**, 161–8.

Hadfield, W. (1968) Leaf temperature, leaf pose and productivity of the tea bush. *Nature, London* **219**, 282–4.

Hadfield, W. (1974) Shade in north-east Indian tea plantations. I. The shade pattern. *Journal of Applied Ecology* **11**, 151–78.

Hadley, P., Roberts, E. H., Summerfield, R. J. and Minchin, F. R. (1983) A quantitative model of reproductive development in cowpea (*Vigna unguiculata* (L.) Walp.) in relation to photoperiod and temperature, and implications for screening germplasm. *Annals of Botany* **51**, 531–43.

Hadley, P., Roberts, E. H., Summerfield, R. J. and Minchin, F. R. (1984) Effects of temperature and photoperiod on flowering in soybean (*Glycine max* (L.) Merrill.): a quantitative model. *Annals of Botany* **53**, 669–81.

Hall, A. E., Foster, K. W. and Waines, J. G. (1979) Crop adaptation to semi-arid environments. In: Hall, A. E., Cannell, G. H. and Lawton, H. W. (eds) *Ecological Studies. Agriculture in Semi-Arid Environments.* Springer-Verlag, Berlin, pp. 148–79.

Hamdi, Q. A., Harris, D. and Clark, J. A. (1987) Saturation deficit, canopy formation and function in *Sorghum bicolor. Journal of Experimental Botany* **38**, 1272–83.

Harris, D., Hamdi, Q. A. and Terry, A. C. (1987) Germination and emergence of *Sorghum bicolor*: genotypic and environmentally induced variation in the response to temperature and depth of sowing. *Plant, Cell and Environment* **10**, 501–8.

Harris, D., Matthews, R. B., Nageswara Rao, R. C. and Williams, J. H. (1988) The physiological basis for yield differences between four genotypes of groundnut (*Arachis hypogaea*) in response to drought. III. Developmental processes. *Experimental Agriculture* **24**, 215–26.

Hawkins, R. C. and Cooper, P. J. M. (1981) Growth, development and grain yield of maize. *Experimental Agriculture* **17**, 203–7.

Hayahsi, K. and Ito, H. (1962) Studies on form of plant in rice varieties with particular reference to efficiency in utilising sunlight. I. The significance of extinction coefficient in rice plant communities. *Proceedings of the Crop Science Society of Japan* **30**, 329–33.

Hesketh, J. D., Baker, D. N. and Duncan, W. G. (1972) Simulation of growth and yield in cotton. II. Environmental control of morphogenesis. *Crop Science* **12**, 436–9.

Howeler, R. H. and Cadavid, L. F. (1983) Accumulation and distribution of dry matter and nutrients during a 12-month growth cycle of cassava. *Field Crops Research* **7**, 123–39.

Hughes, G. and Keatinge, J. D. H. (1983) Solar radiation interception, dry matter production and yield in pigeon pea(*Cajanus cajan* (L.) Millspaugh). *Field Crops Research* **6**, 171–8.

Hughes, G., Keatinge, J. D. H. and Scott, S. P. (1980) Planting density effects on the dry season productivity of short pigeonpeas in the West Indies: 1. Growth and development. *Proceedings of International Workshop on Pigeonpeas*, pp. 235–40. ICRISAT, Patancheru P.O., Andhra Pradesh 502324, India.

Hughes, G., Keatinge, J. D. H. and Scott, S. P. (1981) Pigeon pea as a dry season crop in Trinidad, West Indies. II. Interception and utilization of solar radiation. *Tropical Agriculture (Trinidad)* **58**, 191–9.

Hughes, G., Keatinge, J. D. H., Cooper, P. J. M. and Dee, N. F. (1987) Solar radiation interception and utilization by chickpea (*Cicer arietinum* L.) crops in northern Syria. *Journal of Agricultural Science, Cambridge* **108**, 419–24.

Irikura, Y., Cock, J. H. and Kawano, K. (1979) The physiological basis of genotype-temperature interactions in cassava. *Field Crops Research* **2**, 227–39.

Jones, E., Nyamudesa, P. and Busangavanye, T. (in press). Rainfed cropping and water conservation and concentration on vertisols in the south east lowveld of Zimbabwe. *Proceedings of a Workshop on Vertisol Management in Africa, in Harare, Zimbabwe, January 1979*. International Board for Soil Research and Management, Bangkok, Thailand.

Kasanga, H. and Monsi, M. (1954) On the light transmission of leaves, and its meaning for the production of dry matter in plant communities. *Japan Journal of Botany* 14, 304–24.

Kassam, A. H. and Andrews, D. J. (1975) Effects of sowing date on growth, development and yield of photosensitive sorghum at Samaru, northern Nigeria. *Experimental Agriculture* 11, 227–40.

Kassam, A. H. and Kowal, J. M. (1975) Water use, energy balance and growth of gero millet at Samaru, northern Nigeria. *Agricultural Meteorology* 15, 333–42.

Kassam, A. H., Kowal, J. M. and Harkness, C. (1975) Water use and growth of groundnut at Samaru, northern Nigeria. *Tropical Agriculture* 52, 105–12.

Keating, B. A. and Evenson, J. P. (1979) Effect of soil temperature on sprouting and sprout elongation of stem cuttings of cassava (*Manihot esculenta* Crantz). *Field Crops Research* 2, 241–51.

Kowal, J. M. and Kassam, A. H. (1973) Water use, energy balance and growth of maize at Samaru, northern Nigeria. *Agricultural Meteorology* 12, 391–406.

Lemcoff, J. H. and Loomis, R. S. (1986) Nitrogen influences on yield determination in maize. *Crop Science* 26, 1017–22.

Leong, S. K. and Ong, C. K. (1983) The influence of temperature and soil water deficit on the development and morphology of groundnut (*Arachis hypogaea*). *Journal of Experimental Botany* 34, 1551–61.

Littleton, E. J., Dennett, M. D., Elston, J. and Monteith, J. L. (1979a) The growth and development of cowpeas (*Vigna unguiculata*) under tropical field conditions. 1. Leaf area. *Journal of Agricultural Science, Cambridge* 93, 291–307.

Littleton, E. J., Dennett, M. D., Monteith, J. L. and Elston, J. (1979b) The growth and development of cowpeas (*Vigna unguiculata*) under tropical field conditions. 2. Accumulation and partition of dry weight. *Journal of Agricultural Science, Cambridge* 93, 309–20.

Littleton, E. J., Dennett, M. D., Elston, J. and Monteith, J. L. (1981) The growth and development of cowpeas (*Vigna unguiculata*) under tropical field conditions. 3. Photosynthesis of leaves and pods. *Journal of Agricultural Science, Cambridge* 97, 539–50.

Lugg, D. G. and Sinclair, T. R. (1981) Seasonal changes in photosynthesis of field-grown soybean leaflets. 2. Relation to nitrogen content. *Photosynthesis* 15, 138–44.

McCulloch, J. S. G., Pereira, H. C., Kerfoot, O. and Goodchild, N. A. (1965) Effect of shade trees on tea yields. *Agricultural Meteorology* 2, 385–99.

McRee, K. J. (1964) Equations for the rate of dark respiration of white clover and grain sorghum, as functions of dry weight, photosynthetic rate and temperature. *Crop Science* 14, 509–14.

Mahalakshmi, V. and Bidinger, F. R. (1985a) Flowering response of pearl millet to water stress during panicle development. *Annals of Applied Biology* 106, 571–8.

Mahalakshmi, V. and Bidinger, F. R. (1985b) Water stress and time of floral initiation in pearl millet. *Journal of Agricultural Science, Cambridge* 105, 437–45.

Mahalakshmi, V. and Bidinger, F. R. (1986) Water deficit during panicle development on pearl millet: yield compensation by tillers. *Journal of Agricultural Science, Cambridge* 106, 113–9.

Mahalakshmi, V., Bidinger, F. R. and Raju, D. S. (1987) Effect of timing of water deficit on pearl millet (*Pennisetum americanum*). *Field Crops Research* **15**, 327–39.

Marshall, B. and Willey, R. W. (1983) Radiation interception and growth in an intercrop of pearl millet/groundnut. *Field Crops Research* **7**, 141–60.

Matthews, R. B., Harris, D., Nageswara Rao, R. C., Williams, J. H. and Wadia, K. D. R. (1988a) The physiological basis for yield differences between four genotypes of groundnut (*Arachis hypogaea*) in response to drought. I. Dry matter production and water use. *Experimental Agriculture* **24**, 191–202.

Matthews, R. B., Harris, D., Williams, J. H. and Nageswara Rao, R. C. (1988b) The physiological basis for yield differences between four genotypes of groundnut (*Arachis hypogaea*) in response to drought. 2. Solar radiation interception and leaf movement. *Experimental Agriculture* **24**, 203–13.

Matthews, R. B., Azam-Ali, S. N. and Peacock, J. M. (1990a) Response of four sorghum genotypes to mid-season drought. 2. Leaf characteristics and solar radiation interception. *Field Crops Research* (in press).

Matthews, R. B., Reddy, D. M., Rani, A. U. and Azam-Ali, S. N. (1990b) Response of four sorghum genotypes to mid-season drought. 1. Dry matter production, yield and water use. *Field Crops Research* (in press).

Mohamed, H. A., Clark, J. A. and Ong, C. K. (1985) The influence of temperature during seed development on the germination characteristics of millet seeds. *Plant, Cell and Environment* **8**, 361–2.

Mohamed, H. A., Clark, J. A. and Ong, C. K. (1988a) Genotypic differences in the temperature responses of tropical crops. I. Germination characteristics of groundnut (*Arachis hypogaea* L.) and pearl millet (*Pennisetum typhoides* S. &. H.). *Journal of Experimental Botany* **39**, 1121–8.

Mohamed, H. A., Clark, J. A. and Ong, C. K. (1988b) Genotypic differences in the temperature responses of tropical crops. II. Seedling emergence and leaf growth of groundnut (*Arachis hypogaea* L.) and pearl millet (*Pennisetum typhoides* S. & H.). *Journal of Experimental Botany* **39**, 1129–35.

Mohamed, H. A., Clark, J. A. and Ong, C. K. (1988c) Genotypic differences in the temperature responses of tropical crops. III. Light interception and dry matter production of pearl millet (*Pennisetum typhoides* S. &. H.). *Journal of Experimental Botany* **39**, 1137–43.

Monteith, J. L. (1972) Solar radiation and productivity in tropical ecosystems. *Journal of Applied Ecology* **9**, 747–66.

Monteith, J. L. (1978) Reassessment of maximum growth rates for C3 and C4 crops. *Experimental Agriculture* **14**, 1–5.

Monteith, J. L. (1981) Evaporation and surface temperature. *Quarterly Journal of the Royal Meteorological Society* **107**, 1–27.

Monteith, J. L. (1986a) Significance of the coupling between saturation vapour pressure deficit and rainfall in monsoon climates. *Experimental Agriculture* **22**, 329–38.

Monteith, J. L. (1986b) How do crops manipulate water supply and demand? *Philosophical Transactions of the Royal Society, London* A **316**, 245–59.

Monteith, J. L. and Elston, J. (1983) Performance and productivity of foliage in the field. In: Dale, J. E. and Milthorpe, F. L. (eds) *The Growth and Functioning of Leaves*. Cambridge University Press, Cambridge, pp. 499–518.

Muchow, R. C. (1985) An analysis of the effects of water deficits on grain legumes grown in a semi-arid tropical environment in terms of radiation interception and its efficiency of use. *Field Crops Research* **11**, 309–23.

Muchow, R. C. (1988a) Effect of nitrogen supply on the comparative productivity of maize and sorghum in a semi-arid tropical environment. 1. Leaf growth and leaf nitrogen. *Field Crops Research* **18**, 1–16.

Muchow, R. C. (1988b) Effect of nitrogen supply on the comparative productivity of maize and sorghum in a semi-arid tropical environment. 3. Grain yield and nitrogen accumulation. *Field Crops Research* **18**, 31–43.

Muchow, R. C. and Davis, R. (1988) Effect of nitrogen supply on the comparative productivity of maize and sorghum in a semi-arid tropical environment. 2. Radiation interception and biomass accumulation. *Field Crops Research* **18**, 17–30.

Murata, Y. (1969) Physiological responses to nitrogen in plants. In: Eastin, J. D., Haskins, F. A., Sullivan, C. Y. and Van Bavel, C. H. M. (eds) *Physiological Aspects of Crop Yield*, American Society of Agronomy, Madison, Wisconsin, USA, pp. 235–63.

Murray, F. W. (1967) Computation of saturation vapour pressure. *Journal of Applied Meteorology* **6**, 203–4.

Nambiar, P. T. C., Rego, T. J. and Srinivasa Rao, B. (1986) Comparison of the requirements and utilization of nitrogen by genotypes of sorghum (*Sorghum bicolor* (L.) Moench), and nodulating and non-nodulating groundnut (*Arachis hypogea* L.). *Field Crops Research* **15**, 165–79.

Natarajan, M. and Willey, R. W. (1980a) Sorghum–pigeonpea intercropping and the effects of plant population density. 1. Growth and yield. *Journal of Agricultural Science, Cambridge* **95**, 51–8.

Natarajan, M. and Willey, R. W. (1980b) Sorghum–pigeonpea intercropping and the effects of plant population density. 2. Resource use. *Journal of Agricultural Science, Cambridge* **95**, 59–65.

Natarajan, M. and Willey, R. W. (1986) The effects of water stress on yield advantages of intercropping systems. *Field Crops Research* **13**, 117–31.

Ofori, F. and Stern, W. R. (1987) The combined effects of nitrogen fertilizer and density of the legume component on production efficiency in a maize/cowpea intercrop system. *Field Crops Research* **16**, 43–52.

Ojehomon, O. O. (1970) Effect of continuous removal of open flowers on the seed yield of two varieties of cowpea, *Vigna unguiculata* (L.) Walp. *Journal of Agricultural Science, Cambridge* **74**, 375–81.

Ong, C. K. (1983a) Response to temperature in a stand of pearl millet (*Pennisetum typhoides* S. & H.). 1. Vegetative development. *Journal of Experimental Botany* **34**, 322–36.

Ong, C. K. (1983b) Response to temperature in a stand of pearl millet (*Pennisetum typhoides* S. & H.). 2. Reproductive development. *Journal of Experimental Botany* **34**, 337–48.

Ong, C. K. (1984) The influence of temperature and water deficit on the partitioning of dry matter in groundnut (*Arachis hypogaea* L.). *Journal of Experimental Botany* **35**, 746–55.

Ong, C. K., Black, C. R., Simmonds, L. P. and Saffell, R. A. (1985) Influence of saturation deficit on leaf production and expansion in the stands of groundnut (*Arachis hypogaea*) grown without irrigation. *Annals of Botany* **56**, 528–36.

Ong, C. K. and Monteith, J. L. (1985) Response of pearl millet to light and temperature. *Field Crops Research* **11**, 141–60.

Ong, C. K., Simmonds, L. P. and Matthews, R. B. (1987) Responses to saturation deficit in a stand of groundnut (*Arachis hypogaea* L.). 2. Growth and development. *Annals of Botany* **59**, 121–8.

Ong, C. K. and Squire, G. R. (1984) Response to temperature in a stand of pearl millet (*Pennisetum typhoides* S. & H.). 7. Final numbers of spikelets and grains. *Journal of Experimental Botany* **35**, 1233–40.

Ougham, H. J., Peacock, J. M., Stoddart, J. L. and Soman, P. (1988) High temperature effects on seedling emergence and embryo protein synthesis of sorghum. *Crop Science* **28**, 251–3.

Ougham, H. J. and Stoddart, J. L. (1985) Development of a laboratory screening technique, based on embryo protein synthesis, for the assessment of high-temperature susceptibility during germination of *Sorghum bicolor*. *Experimental Agriculture* **21**, 343–55.

Ougham, H. J. and Stoddart, J. L. (1986) Synthesis of heat-shock protein and acquisition of thermotolerance in high-temperature tolerant and high-temperature susceptible lines of sorghum. *Plant Science* **44**, 163–7.

Pate, J. S. and Minchin, F. R. (1980) Comparative studies of carbon and nitrogen nutrition of selected grain legumes. In: Summerfield, R. J. and Bunting, A. H. (eds) *Advances in Legume Science*, HMSO, London, pp. 105–14.

Penman, H. L. (1948) Natural evaporation from open water, bare soil and grass. *Proceedings of the Royal Society, London* **A193**, 120–45.

Puckridge, D. W. and O'Toole, J. C. (1981) Dry matter and grain production of rice, using a line source sprinkler in drought studies. *Field Crops Research* **3**, 303–19.

Pyare Lal, Mishra, G. N., Gautam, R. C. and Bisht, P. S. (1982) Response of transplanted rice to different plant populations under limited supply of nitrogen. *Indian Journal of Agricultural Science* **52**, 372–7.

Quinby, J. R., Hesketh, J. D. and Voigt, R. L. (1973) Influence of temperature and photoperiod on floral initiation and leaf number in sorghum. *Crop Science* **13**, 243–6.

Rao, R. C. N., Simmonds, L. P., Azam-Ali, S. N. and Williams, J. H. (1989) Population, growth and water use of groundnut maintained on stored water. 1. Root and shoot growth. *Experimental Agriculture* **25**, 51–61.

Rawson, H. M. and Constable, G. A. (1980) Gas exchange of pigeonpea: a comparison with other crops and a model of carbon production and its distribution within the plant. *Proceedings of International Workshop on Pigeonpeas*, **Volume 1**, pp. 175–89. ICRISAT, Patancheru P.O., Andhra Pradesh 502324, India.

Reddy, M. D., Ghosh, B. C. and Panda, M. M. (1986) Effect of seed rate and application of N fertiliser on grain yield and N uptake of rice under intermediate deepwater conditions (15–50 cm). *Journal of Agricultural Science, Cambridge* **107**, 61–6.

Reddy, M. S. and Willey, R. B. (1981) Growth and resource use studies in an intercrop of pearl millet/groundnut. *Field Crops Research* **4**, 13–24.

Rees, A. R. (1963) Relationship between growth rate and leaf area index in the oil palm. *Nature* **197**, 63–4.

Rees, A. R. and Tinker, P. B. H. (1963) Dry matter production and nutrient content of plantation oil palms in Nigeria. 1. Growth and dry matter production. *Plant and Soil* **19**, 19–31.

Rees, D. J. (1986a) Crop growth, development and yield in semi-arid conditions in Botswana. I. The effects of population density and row spacing on *Sorghum bicolor*. *Experimental Agriculture* **22**, 153–67.

Rees, D. J. (1986b) The effects of population density, row spacing and intercropping on the interception and utilisation of solar radiation by *Sorghum bicolor* and *Vigna unguiculata* in semi-arid conditions in Botswana. *Journal of Applied Ecology* **23**, 917–28.

Rees, D. J. (1986c) The effects of population density and intercropping with cowpea on the water use and growth of sorghum in semi-arid conditions in Botswana. *Agricultural and Forest Meteorology* **37**, 293–308.

Rijks, D. A. (1971) Water use by irrigated cotton in the Sudan. III. Bowen ratios and advective energy. *Journal of Applied Ecology* **8**, 643–63.

Rijks, D. A. (1976) Water use by irrigated cotton in the Sudan. IV. Water use, potential evaporation and yield. *Journal of Applied Ecology* **13**, 491–506.

Ripley, E. A. (1967) Effects of shade and shelter on the microclimate of tea and their significance. *East African Agriculture and Forestry Journal* **33**, 67–80.

Ritchie, J. T. (1972) Model for predicting evaporation from a row crop with incomplete cover. *Water Resources Research* **8**, 1204–11.

Rowden, R., Gardiner, D., Whiteman, P. C. and Wallis, E. S. (1981) Effects of planting density on growth, light interception and yield of a photoperiod insensitive pigeon pea (*Cajanus cajan*). *Field Crops Research* **4**, 201–13.

Sardar Singh and Russell, M. B. (1980) Water use by a maize/pigeon pea intercrop on a deep vertisol. *Proceedings of International Workshop on Pigeon Peas*, **Volume 1**, pp. 271–82. ICRISAT, Patancheru P.O., Andhra Pradesh 502324, India.

Satake, T. (1976) Sterile type cool injury in paddy rice plants. In: *Climate and Rice*, IRRI, Los Banos, Philippines, pp. 281–300.

Shackell, K. A. and Hall, A. E. (1979) Reversible leaf movements in relation to drought adaptation of cowpeas, *Vigna unguiculata* (L.) Walp. *Australian Journal of Plant Physiology* **6**, 265–76.

Sheldrake, A. R. and Narayanan, A. (1979) Growth, development and nutrient uptake in pigeonpeas (*Cajanus cajan*). *Journal of Agricultural Science, Cambridge* **92**, 513–26.

Sheldrake, A. R., Narayanan, A. and Venkataratnam, N. (1979) The effects of flower removal on the seed yield of pigeonpeas (*Cajanus cajan*). *Annals of Applied Biology* **91**, 383–90.

Simmonds, L. P. and Azam-Ali, S. N. (1989) Population, growth and water use of groundnut maintained on stored water. 4. The influence of population on water supply and demand. *Experimental Agriculture* **25**, 87–98.

Simmonds, L. P. and Ong, C. K. (1987) Responses to saturation deficit in a stand of groundnut (*Arachis hypogaea* L.) 1. Water use. *Annals of Botany* **59**, 113–9.

Simmonds, L. P. and Williams, J. H. (1989) Population, water use and growth of groundnut maintained on stored water. 2. Transpiration and evaporation from soil. *Experimental Agriculture* **25**, 63–75.

Sinclair, T. R. and de Wit, C. T. (1976) Analysis of the carbon and nitrogen limitations to soybean yield. *Agronomy Journal* **68**, 319–24.

Sivakumar, M. V. K. and Virmani, S. M. (1980) Growth and resource use of maize, pigeonpea and maize/pigeonpea intercrop in an operational research watershed. *Experimental Agriculture* **16**, 377–86.

Snyder, R. L. (1985) Hand calculating degree days. *Agricultural and Forest Meteorology* **35**, 353–8.

Soman, P. and Peacock, J. M. (1985) A laboratory technique to screen seedling emergence of sorghum and pearl millet at high soil temperature. *Experimental Agriculture* **21**, 335–41.

Squire, G. R. (1979) Weather, physiology and seasonality of tea (*Camellia sinensis*) yields in Malawi. *Experimental Agriculture* **15**, 321–30.

Squire, G. R. (1986) A physiological analysis for oil palm trials. *Porim Bulletin* **12**, 12–31, (Palm Oil Research Institute of Malaysia, Kuala Lumpur).

Squire, G. R. (1989a) Responses to temperature in a stand of pearl millet. 9. Expansion processes. *Journal of Experimental Botany* **40**, 1383–9.

Squire, G. R. (1989b) Responses to temperature in a stand of pearl millet. 10. Partition of assimilate. *Journal of Experimental Botany* **40**, 1391–8.

Squire, G. R., Black, C. R. and Ong, C. K. (1983) Response to saturation deficit of leaf extension in a stand of pearl millet (*Pennisetum typhoides* S. & H.). 2. Dependence on leaf water status and irradiance. *Journal of Experimental Botany* **34**, 856–65.

Squire, G. R. and Callander, B. A. (1981) Tea Plantations. In: Kozlowski, T. T. (ed.) *Water Deficits and Plant Growth, VI*, Academic Press, New York, pp. 471–510.

Squire, G. R., Gregory, P. J., Monteith, J. L., Russell, M. B. and Piara Singh. (1984) Control of water use by pearl millet (*Pennisetum typhoides* S. & H.). *Experimental Agriculture* **20**, 135–9.

Squire, G. R. and Ong, C. K. (1983) Response to saturation deficit of leaf extension in a stand of pearl millet (*Pennisetum typhoides* S. & H.). 1. Interaction with temperature. *Journal of Experimental Botany* **34**, 846–55.

Stephens, W. and Carr, M. K. V. (1990) Seasonal and clonal differences in shoot extension rates and numbers in tea (*Camellia sinensis* L.). *Experimental Agriculture* (in press).

Stirling, C. M., Ong, C. K. and Black, C. R. (1989) The response of groundnut (*Arachis hypogaea* L.) to timing of irrigation. 1. Development and growth. *Journal of Experimental Botany* **40**, 1145–53.

Stirling, C. M., Williams, J. H., Black, C. R. and Ong, C. K. (1990) The effect of timing of shade on development, dry matter production and light-use efficiency in groundnut (*Arachis hypogaea* L.) under field conditions. *Australian Journal of Agricultural Science* **41** (in press).

Tanner, C. B. and Jury, W. A. (1976) Estimating evaporation and transpiration from a row crop during incomplete cover. *Agronomy Journal* **68**, 239–43.

Tanner, C. B. and Sinclair, T. R. (1983) Efficient water use in crop production. In: Taylor, H. M., Jordan, W. R. and Sinclair, T. R. (eds) *Limitations to Efficient Water Use*, American Society of Agronomy, Madison, Wisconsin, USA, pp. 1–27.

Tanton, T. W. (1979) Some factors limiting yields of tea (*Camellia sinensis*). *Experimental Agriculture* **15**, 187–91.

Tanton, T. W. (1981) Growth and yield of the tea bush. *Experimental Agriculture* **17**, 323–31.

Tanton, T. W. (1982a) Environmental factors affecting the yield of tea (*Camellia sinensis*). I. Effects of air temperature. *Experimental Agriculture* **18**, 47–52.

Tanton, T. W. (1982b) Environmental factors affecting the yield of tea (*Camellia sinensis*). II. Effects of soil temperature, daylength and dry air. *Experimental Agriculture* **18**, 53–63.

Tayo, T. O. (1982) Growth, development and yield of pigeon pea (*Cajanus cajan* (L.) Millsp.) in the lowland tropics. 1. Effect of plant population density. *Journal of Agricultural Science, Cambridge* **98**, 65–9.

Thom, A. S. and Oliver, H. R. (1977) On Penman's equation for estimating regional evaporation. *Quarterly Journal of the Royal Meteorological Society* **103**, 345–57.

Turk, K. J., Hall, A. E. and Asbell, C. W. (1980a) Drought adaptation of cowpea. 2. Influence of drought on plant water status and relations with seed yield. *Agronomy Journal* **72**, 421–7.

Turk, K. J. and Hall, A. E. (1980b) Drought adaptation of cowpea. 3. Influence of drought on plant growth and relations with seed yield. *Agronomy Journal* **72**, 428–33.

Turk, K. J., Hall, A. E. and Asbell, C. W. (1980c) Drought adaptation of cowpea. 1. Influence of drought on seed yield. *Agronomy Journal* **72**, 413–20.

Turner, N. C., O'Toole, J. C., Cruz, R. T., Namuco, O. S. and Ahmad, S. (1986a) Responses of seven diverse rice cultivars to water deficits. 1. Stress development, canopy temperature, leaf rolling and growth. *Field Crops Research* **13**, 257–71.

Turner, N. C., O'Toole, J. C., Cruz, R. T., Yambao, E. B., Ahmad, S., Namuco, O. S. and Dingkuhn, M. (1986b) Responses of seven diverse rice cultivars to water deficits. 2. Osmotic adjustment, leaf elasticity, leaf extension, leaf death, stomatal conductance and photosynthesis. *Field Crops Research* **13**, 273–86.

Veltkamp, H. J. (1985) Physiological causes of yield variation in cassava (*Manihot esculenta* Crantz). *Agricultural University Wageningen Papers* **85–6**. Agricultural University, Wageningen, The Netherlands.

Venkataratnam, N. and Green, J. M. (1979) *Pulse Physiology Progress Report 1978–1979, Part 1: Pigeonpea physiology*, pp. 66–9. ICRISAT, Patancheru P.O., Andhra Pradesh 502324, India.

Vergara, B. S. and Chang, T. T. (1976) *The Flowering Response of the Rice Plant to Photoperiod*. International Rice Research Institute, Los Banos, Philippines, 75 pp.

Wallace, J. S., Batchelor, C. H., Dabeesing, D. N. and Soopramanien, G. C. (1990) The partitioning of light and water in drip irrigated plant cane with a maize intercrop. *Agricultural Water Management* (in press).

Wardlaw, I. F., Sofield, I. and Cartwright, P. M. (1980) Factors limiting the rate of dry matter accumulation in the grain of wheat grown at high temperature. *Australian Journal of Plant Physiology* **7**, 387–400.

Warren Wilson, J. (1967) Ecological data on dry matter production. In: Bradley, E. F. and Denmead, O. T. (eds) *The Collection and Processing of Field Data*, Interscience, New York, pp. 77–123.

Westlake, D. F. (1963) Comparisons of plant productivity. *Biological Reviews* **38**, 385–425.

Wien, H. C., Littleton, E. J. and Ayanaba, A. (1979) Drought stress of cowpea and soybean under tropical conditions. In: Mussell, H. and Staples, R. C. (eds) *Stress Physiology in Crop Plants*, Wiley Interscience, New York, pp. 283–301.

Wien, H. C. and Summerfield, R. J. (1980) Adaptation of cowpeas in West Africa: effects of photoperiod and temperature responses in cultivars of diverse origin. In: Summerfield, R. J. and Bunting, A. H. (eds) *Advances in Legume Science*, HMSO, London, pp. 405–17.

Williams, J. B. (1979) *Physical aspects of water use under traditional and modern irrigation/farming systems in the Wadi Rima' Tihamah*. Land Resources Development Centre, Overseas Development Natural Resources Institute, Central Avenue, Chatham Maritime, Chatham, Kent, U.K.

Williams, J. H., Wilson, J. H. H. and Bate, G. C. (1975) The growth of groundnuts (*Arachis hypogaea* L. cv. Makulu red) at three altitudes in Rhodesia. *Rhodesian Journal of Agricultural Research* **13**, 33–43.

Williams, W. A., Loomis, R. S. and Lepley, C. R. (1965) Vegetative growth of corn as affected by population density. I. Productivity in relation to interception of solar radiation. *Crop Science* **5**, 211–15.

Yoshida, S. (1977) Rice. In: Alvim, P. de T. and Kozlowski, T. T. (eds) *Ecophysiology of Tropical Crops*, Academic Press, New York, pp. 857–87.

Yoshida, S. and Hasegawa, S. (1982) In: *Drought Resistance in Crops with Emphasis on Rice*. International Rice Research Institute, Los Banos, Philippines.

Yoshida, S. and Parao, F. T. (1976) Climatic influence on yield and yield components of lowland rice in the tropics. In: *Climate and Rice*, International Rice Research Institute, Los Banos, Philippines, pp. 471–94.

Index